The Right(s) to Water

Pierre Thielbörger

The Right(s) to Water

The Multi-Level Governance of a Unique Human Right

Springer

Pierre Thielbörger
Institute for International Law of Peace
and Armed Conflict
Ruhr-Universität Bochum
Bochum
Germany

ISBN 978-3-642-33907-3 ISBN 978-3-642-33908-0 (eBook)
DOI 10.1007/978-3-642-33908-0
Springer Heidelberg New York Dordrecht London

Library of Congress Control Number: 2013948394

Printed on acid-free paper

Springer is part of Springer Science+Business Media (www.springer.com)

Abstract

As indicated in the title "The Right(s) to Water", this book argues that one single self-standing, comprehensive, and legally binding human right to water is not easily found in international law. However, it is important to note that this is not at all the same thing as stating that "there is no human right to water". It is simply to say that this human right does not easily meet some of the characteristics of other human rights all at once. It is rather a right of its very own kind that must be seen in connection with national guarantees with regard to water and with other recognized human rights, such as the right to life, the right to the highest attainable standard of health, and the right to an adequate standard of living.

The book has three main chapters. A first analytical part examines the present status of the right to water in international, European, and domestic laws. A second, theoretical part scrutinizes whether and in which form it is conceptually best possible to consider water as the object of a distinct human right. Finally, a third, applied part considers how concrete reforms and developments can improve the effective implementation of the right in practical terms. Only by considering the right to water in its legal, philosophical, and practical context can the present status and future potential of the right to water be fully captured.

Acknowledgements

I have written this book over the course of the last years in many wonderful places, among them, foremost, Florence and Boston, as well as Maastricht, Bochum, and Berlin. In all these places, I was lucky to find colleagues and friends who have supported me and my academic work with great passion.

In Florence, Italy, during the years of the PhD my *alma mater*, I would particularly like to thank Professor Ernst-Ulrich Petersmann. Although he was Head of the Law Department of the European University Institute for many of the last years, he always found time to discuss my ideas with me. He was a great source of advice, support, and knowledge. I am, and remain, deeply indebted to him. I would also like to thank Professor Martin Scheinin, whose great knowledge in the field of human rights was hugely helpful for me. He was always interested in my ideas and willing to facilitate contacts to other academics working in the field.

For my time in Boston, USA, I would like to thank Professor Philip Alston, now at New York University but back then at Harvard Law School. I have always found his work very inspirational from the very first days of my legal studies. He has influenced the research on economic and social rights more than anybody else, and I was very lucky to discuss my ideas with and find a strong supporter in him. I would also like to thank Professor Mathias Risse from Harvard Kennedy School, and with him the Harvard Kennedy School Carr Center for Human Rights Policy. Professor Risse and his writings awoke my interest in philosophical approaches to global inequalities and human rights. I am deeply thankful to him.

I would also like to thank Catarina de Albuquerque, United Nations Special Rapporteur on the human right to water and sanitation. I was very glad she found time in her busy schedule to discuss her mandate with me, which allowed my work to include practical considerations with regard to implementing the right human right to water. I wish her all the best for the fulfillment of her important mandate and look forward to a continued cooperation in the future.

On a more personal note, I would like to thank my colleagues and friends who supported me over the last years in Florence, Boston, Maastricht, Bochum, and Berlin. Just to name a few, I would like to thank in particular Tobias Ackermann,

Sören Affeldt, Stan van Alphen, Anna Beckers, Kathrin Bimesdörfer, Nuhaila Carmouche, Sacha Garben, Kathrin Hamenstädt, Maike Heße, Felicitas Kaape, Dilan Khoshnaw and Adibeli Nduka-Agwu.

These cities—Florence, Boston, Maastricht, Bochum, and Berlin—were not only workplaces but homes at the same time. I hope it will continue to feel like home when I return to them in the future.

More than anybody else, I would like to thank my family. Mark has supported me in the course of the last years more than anybody else, and deserves very special thanks. I am also thankful to Nicole and to my mother for their support over the last years. I know my father would have supported me just as much, if it only had been possible. It is to him that this PhD is dedicated.

Abbreviations

AA	Auswärtiges Amt (Federal Foreign Office of Germany)
AdA	Compania de Aguas del Aconquija
AfCRC	African Charter on the Rights and Welfare of the Child
AIDS	Acquired immunodeficiency syndrome
AIR	All India Reporter
BCLR	Butterworths Constitutional Law Reports
BerlBG	Berliner Betriebe-Gesetz
BGH	Bundesgerichtshof (Federal Court of Justice of Germany)
BGHZ	Entscheidungssammlung des Bundesgerichtshof in Zivilsachen (Collected decisions of the Federal Court of Justice of Germany)
BIT	Bilateral Investment Treaty
BMU	Bundesministerium für Umwelt, Naturschutz und Reaktorsicherheit (Federal Ministry of the Environment, Nature Conservation and Nuclear Safety)
BVerfG	Bundesverfassungsgericht (Federal Constitutional Court of Germany)
BVerfGE	Bundesverfassungsgericht Entscheidungen (Collected decisions of the Federal Constitutional Court of Germany)
BVerwG	Bundesverwaltungsgericht (Federal Administrative Court of Germany)
CCPR	Committee Human Rights Committee
CFDD	Conseil Fédéral du Développement Durable
CFI	Court of First Instance
CFREU	Charter of Fundamental Rights of the European Union
CGCT	Code général des collectivés territoriales
CGE	Compagnie Générale des Eaux
COHRE	Centre on Housing Rights and Evictions
CRC	Convention on the Rights of the Child
DFID	Department for International Development
EC	European Community

ECHR	European Convention for the Protection of Human Rights and Fundamental Freedoms
ECJ	European Court of Justice
ECOSOC	United Nations Economic and Social Council
ECSR	European Committee of Social Rights
ESC	European Social Charter
ESCR-Committee	Committee on Economic, Social and Cultural Rights
EU	European Union
GemO BW	Gemeindeordnung für Baden-Württemberg
HIV	Human immunodeficiency virus
HRC	Human Rights Council
ICCPR	International Covenant on Civil and Political Rights
ICESCR	International Covenant on Economic, Social and Cultural Rights
ICJ	International Court of Justice
ICSID	International Centre for Settlement of Investment Disputes
ICSID-Convention	Convention on the Settlement of Investment Disputes between States and Nationals of other States
IDEA	International Institute for Democracy and Electoral Assistance
IFI	International Financial Institution
Ker	Kerala
LEMA	Loi sur l'eau et les milieus aquatiques (French Law on Water and Aquatic Environments)
MDG	Millennium Development Goal
NGO	Non-governmental organization
OAU	Organization of African Unity
OECD	Organisation for Economic Co-operation and Development
OHCHR	Office of the High Commissioner for Human Rights
OJ	Official Journal of the European Communities
PCIJ	Permanent Court of International Justice
SA	South African Law Reports
SCR	Supreme Court Reporter
SEMAPA	Sociedade de Investimento e Gestão
TEU	Treaty on the European Union
TFEU	Treaty on the Functioning of the European Union
TGI	Tribunal de Grande Instance
UDHR	Universal Declaration of Human Rights
UK	United Kingdom
UN	United Nations
UNCLOS	United Nations Convention on the Law of the Sea
UNDP	United Nations Development Programme
UNESCO	United Nations Educational, Scientific and Cultural Organization
UNGA	United Nations General Assembly

UNHCR	United Nations High Commissioner for Refugees
UNICEF	United Nations Children's Fund
UNSC	United Nations Security Council
UNTS	United Nations Treaty Series
UPLBEC	Uttar Pradesh Local Bodies & Educational Cases
UPR	Universal Periodic Review
US or USA	United States of America
VCLT	Vienna Convention on the Law of Treaties
WHG	Wasserhaushaltsgesetz
WHO	World Health Organization
WTO	World Trade Organization

Contents

Chapter 1
Introduction

Water is life. Lack of it leads to illness, misery and death. It is a grave fact that according to reports of the United Nations (UN)[1] in 2012 more than 780 million people rely on unsafe potable water and another 2.5 billion people have no access to sanitation.[2] Almost two million children die each year for want of clean water and clean toilets.[3] A recent example of this water crisis is the outbreak of cholera in 2008 in Zimbabwe, spread by water contaminated with human excrement, leaving the population of the whole country without clean and safe water for several weeks.[4] Another dramatic example is that of Haiti. The initial disaster, the earthquake of 12 January 2010, was followed by a devastating water crisis. Tremors from the initial earthquake demolished the water pipes, allowing seawater to seep into the drinking water supplies and rats to multiply in the sewer systems, making Haiti's freshwater undrinkable for weeks.[5] With no other water sources available, many Haitians nevertheless consumed the contaminated drinking water for weeks, causing a dramatic cholera epidemic.[6] As a third example, South Pacific Island States were confronted with severe water shortages as a consequence of low rainfall

*All cited webpages were last accessed on 1 August 2013.

[1] UNICEF and WHO (2012); WHO (2012).

[2] See similar figures in UN (2009), p. 46, which states that 884 million people worldwide still rely on unimproved water sources for drinking. Following DFID (2001), p. 13, even more than one billion people lack access to safe water. See also UNDP (2006).

[3] See UNDP (2006), pp. 5–6; "Lack of Toilets Among Leading Causes of Childs Death, Says UNICEF on World Toilet Day", in: *UNICEF-USA*, 19 November 2012, http://www.unicefusa.org/news/releases/lack-of-toilets-causes-child-deaths.html.

[4] See for instance Dugger (2008).

[5] See for a good overview of the earthquake and its implications for the water supply in Haiti the reports at the *New York Times*, http://www.nytimes.com/interactive/world/americas/2010-haiti-shattered-year.html#intro.

[6] See for a good overview on the outbreak of cholera on Haiti McNeil Jr (2010) or "More than 100 dead in suspected cholera outbreak in Haiti", in: *The Guardian*, 22 October 2010, http://www.guardian.co.uk/world/2010/oct/22/haiti-cholera-outbreak. See, more recently, Sontag (2012).

P. Thielbörger, *The Right(s) to Water*,
DOI 10.1007/978-3-642-33908-0_1, © Springer-Verlag Berlin Heidelberg 2014

in 2011. Rising sea levels exacerbated the crisis by contaminating the groundwater. As a consequence, the inhabitants of Tuvalu and Tokelau heavily depended on foreign aid in terms of bottled water supply.[7] These examples show that, despite remarkable advances in some areas of the world in order to meet the UN Millennium Development Goals,[8] the global water crisis[9] is evident, in particular amongst the poor rural population.[10] The imperative for the international community to act in order to protect the planet's "blue gold"[11] could not be more urgent.

One way of confronting this global water crisis by reliable legal means is a subjective right to potable water for all humans. In this sense, on 28 July 2010, the UN General Assembly (UNGA) passed a resolution called "The human right to water and sanitation", in which it "recognizes the right to safe and clean drinking water and sanitation as a human right that is essential for the full enjoyment of life and all human rights".[12] A few weeks later, the Human Rights Council (HRC) passed a corresponding resolution affirming that the right "is derived from the right to an adequate standard of living".[13] Despite these affirmations, the question remains: does a human right to water currently exist, to what extent and in what form should it exist at all, and can it make a difference to the global water challenges? These research questions lie at the heart of this book.

As indicated in the title "The Right(s) to Water", I deem it better to look at the right not only as one international human right, but as a complex network of different international and domestic rights. A single international human right does not easily meet some of the characteristics often associated with human rights. Firstly, the common Human Rights Instruments do not recognize a human right to water explicitly; one can only "derive" it from broader human rights recognized therein. Secondly, those legal documents that do recognize the right explicitly do so only for certain groups (such as women, children, etc.), but not comprehensively for

---

[7] See for a good overview the articles "South Pacific islands face water crisis after 6 months of low rainfall", in: *The Guardian*, 4 October 2011, http://www.guardian.co.uk/environment/2011/oct/04/south-pacific-water-crisis-rainfall; "The South Pacific's water crisis: And not a drop to drink", in: *The Economist*, 6 October 2011, http://www.economist.com/blogs/banyan/2011/10/south-pacifics-water-crisis; "Australia, New Zealand in airlift to drought-hit Tuvalu", in: *BBC NEWS*, 7 October 2011, http://www.bbc.co.uk/news/world-asia-pacific-15210568.

[8] As UN Secretary General Ban-Ki Moon states, "we are well on our way to meeting the target for safe drinking water" (UN (2009), p. 3). However, a closer look reveals that the improved statistics are mainly thanks to regional efforts, in particular of China and India, through the increased use of improved water supplies and the expansion of access.

[9] Despite the recent progress, UN (2009) still uses this term, see for instance Sha Zukang, Under-Secretary-General for Economic and Social Affairs, p. 5.

[10] See UN (2012), p. 53, according to which the number of people without an improved water source in rural areas are five times greater than in urban areas.

[11] The term "Blue Gold" was prominently introduced by Barlow and Clarke (2002), and is becoming more and more common in the current debate, see for instance: Petrova (2006) or Kessler (2007).

[12] UNGA, *The human right to water and sanitation*, A/RES/64/292, 3 August 2010, op. par. 1.

[13] HRC, *Human rights and access to safe drinking water and sanitation*, A/HRC/RES/15/9, 6 October 2010, op. par. 3.

all humans. Thirdly, all explicit recognitions that aim at granting the right to all human beings (e.g. the two aforementioned resolutions of the UNGA and the HRC) are not *per se* legally binding. Thus, what is often called "the human right to water" exists only as a right of its very own kind; it is also not clear whether it exists rather as part of treaty law or customary law, or both. It is therefore better to understand the right to water as a complex of different rights. This includes recognitions in domestic constitutional and sub-constitutional law as well the right's close connections to, if not origins in, other accepted human rights.

Although often perceived as a purely socio-economic right, the human right to water, more than other rights, combines components of different legal character. As water is one of the most essential human needs of all, the right also has an important nucleus relating to the human right to life and a mantle relating to the human rights of an adequate standard of living and the human right to the highest attainable standard of health. This triad of human rights also reflects the most suitable normative content of a right to water: without "availability" of water, the right to life is at stake; if water lacks sufficient "quality", the right to health is challenged; and without physical and economic water "accessibility", an adequate standard of living is not possible. In compounding these different legal mosaics, the components indeed form what can by now legitimately be called a "human right to water". It might be of political significance to call these different individual claims relating to water "one human right". However, one must keep in mind that it takes some legal efforts to come to this conclusion under the umbrella of only one right. By understanding it as a right of different layers, however, deeply rooted in both national rights as well as other international human rights, one can turn necessity into a virtue, thereby maximizing the right's possible effects.

In this multi-layered character, the right inevitably raises questions of the relations of the different legal orders. While cherishing the supremacy of international law, I argue that a human right to water must be understood as upholding international minimum levels of protection, which can be outperformed, but not underperformed by national law.

Whether a human right to water can make a crucial contribution to solve the global water crisis will depend (at least) on three significant factors: its independent monitoring (to identify violations); its enforcement against the private sector (as an actor of increasing importance in the field of water delivery); and its ability to motivate international assistance (without which it will not be possible to realize the human right to water globally).

The book addresses its research question in three steps. A first analytical part examines the present status of the right to water in international, European and domestic law. A second, theoretical part, scrutinizes which conceptual and theoretical objections the right faces and in which form the right is conceptually best understood. Finally, a third, applied part considers which elements are necessary to ensure the right can be meaningful in practical terms. Only by considering the right to water in its legal, theoretical and practical context all at once can the present status and future potential of the right to water be fully addressed.

The first descriptive-analytical chapter examines the current legal status of the right to water in different legal orders. One needs to identify how the law currently stands, before looking for answers as to how it can be developed or re-conceptualized in the future. In order to address this question comprehensively, the book scrutinizes all legal levels—the spheres of domestic, European and international law. The study distinguishes between protection that derives from legislation and protection that is judge-made. Starting from the legal orders of European States such a Germany, France or Belgium, the analysis examines how the legal order of the European Union (EU) and that of the Council of Europe has dealt with water rights. Subsequently, important non-European examples, such as South Africa and India, shed light on alternative approaches to a right to water that are significantly different from the European experience. In a last step, the first chapter scrutinizes in detail the approach to a right to water in international law. How do international treaties make mention of water and how have international judges dealt with the issue of basic water needs? The study takes all sources of international law as mentioned in Art. 38 of the Statute of the International Court of Justice (ICJ)—international conventions, international custom, and the general principles of law as recognized by civilized nations—into account. The book gives special attention to General Comment No. 15[14] named "The right to water", released by the Committee on Economic, Social and Cultural Rights (ESCR-Committee) in 2002, as it is widely considered as ground-breaking and highly influential for the recognition of the right in international law.[15] Altogether, the first chapter seeks to answer two questions: Firstly, to what extent do the different legal orders provide effective protection of a right to water, and secondly, to what extent do they provide protection that can be considered an archetype for the protection of the right to water in other legal orders?

The second chapter is theoretical and normative in nature. The chapter addresses general philosophical questions related to a right to water. It seeks to satisfy a number of theoretical objections that can be, and have been, raised against a human right to water (and might be part of the reason why the right took so long to develop). Starting from the very term "human-right to-water", the study examines all three parts of the term. Firstly, is *water* a suitable content for a distinct right? Can we meaningfully hold a right to a limited or scarce resource? Secondly, should we accept as binding upon us a *right* which we are possibly not able to fulfill for all humans? And finally, if one accepts a distinct right to water, would such a right necessarily fulfill the conditions of that very special category of a *human* right?

With these theoretical clarifications in mind, the second chapter raises the question of how best to conceptualize such a right in order to adequately reflect

---

[14] ESCR-Committee, *General Comment No. 15: The right to water (Art. 11 and 12)*, HRI/GEN/1/Rev.9 (Vol. I), 27 May 2008, at 97 (hereinafter "ESCR-Committee, *General Comment No. 15*").

[15] Rudolf argues that the General Comment has decisively influenced ("maßgeblich beeinflusst") the discussion on a right to water ever since its release, see Rudolf (2007), p. 22; see also Langford (2006), p. 479, who argues that the General Comment "has had perhaps more impact and influence than the drafters initially expected."

the different existing approaches. How can the diverse approaches in different legal orders be reconciled? Is the right a self-standing human right or a derivative right? The study sheds light on the question of legal derivation and highlights advantages and disadvantages of this legal technique. For the case of a derivative right, one must identify a legal source: is the right more related to civil and political rights, to economic and social rights, or is a combined approach best reflective of the nature of the right? Should one summarize the spectrum of possible entitlements related to water, ranging from moderate price development in welfare States to access to a minimum of drinking water in developing countries, under the umbrella of one right? And if so, what is the best theoretical architectural framework for such a unified right?

In determining the content of such a right, the study also distinguishes it from another right that is increasingly advocated on the international agenda: the right to sanitation. Should water and sanitation be seen as one human right or are they different in nature and scope?

Finally, the chapter addresses the question of how the recognitions of the right to water on different legal levels relates to each other. The much debated judgments of the European Court of Justice (ECJ) in the case of *Kadi*[16] have shown that the relationship between different legal levels remains highly contentious. It is in cases like the right to water that these distinctions show great practical relevance. Would a human right to water in international law make national recognitions dispensable, or vice versa? And what legal recognition of a right to water prevails, if national or regional law clashes with the international legal order?

With this understanding of the current legal status of the right to water, developed in the first analytical chapter, and with a refined understanding of its normative structure, gained in the second theoretical chapter, one must raise a rather blunt question: what difference can a human right to water make at all? Is there an added value of a rights-based approach to water? The last chapter suggests a focus of the present discussion on the right to water to three specific elements. In order to be effective, the right must be effectively monitored, enforced and realized. The logic behind these elements is the following: Violations must be, in the first place, noticed and recorded ("monitoring"). Secondly, in order to address these violations, the right must be enforced or upheld where it is most crucially challenged, namely against the private sector. And finally, in order to prevent future violations, and address the global water crisis, the right must be progressively realized, not only nationally, but through the means of international water assistance. For these three elements, the study will assess the status quo, identify existing challenges, and, where appropriate, make suggestions for substantive improvement.

With respect to monitoring, the mandate of the new UN Special Rapporteur on the Human Right to Water and Sanitation, formerly called Independent Expert on the issue of human rights obligations related to access to safe drinking water and

[16] ECJ, Joined Cases C-402/05 P and C-415/05 P, *Yassin Abdullah Kadi and Al Barakaat International Foundation v. Council of the European Union and Commission of the European Communities*, Judgment of 3 September 2008, [2008] ECR I-6351.

sanitation,[17] currently held by the Portuguese lawyer Catarina de Albuquerque,[18] deserves special attention. To what extent is the mandate able to provide independent and comprehensive monitoring of violations of a human right to water?

In terms of enforcement towards the private sector, the interplay of a right to water and international investment law assumes a key role. Some see privatization as the silver bullet for the realization of access to water, others condemn it as part of the right's ultimate demise. Thus, how should the private sector best be involved? What are the reasons why privatization in the water sector can work, or why is it doomed to fail, and how does one shift the parameters in order to best realize a human right to water in times of water privatization? This will also include an economic analysis, shedding light on the question of to what extent the two concepts of water as a human right, and its service delivery through the private sector, can go together. Disputes between States and international investors, if such privatization fails, are examined through analyzing specific cases before the International Centre for the Settlement of Investment Disputes (ICSID)—the forum which typically decides over water investment disputes between States and investors. How could a human right to water contribute to upholding the rights of people and ensure the protection of basic needs, while at the same times respecting the legitimate interests of foreign investors?

The final element for the right to be effective is that of international realization. Is there a sound moral and legal case to be made for such water assistance, or is it ultimately an anchorless claim of the developing world? What are the problems that still hold back international water assistance, and how, if at all, can a right to water be helpful to overcome them?

Much research has, of course, been done on an important topic like that of a human right to water. Although this research is highly valuable and has advanced the discussion on a right to water, there are still significant gaps that this book strives to fill. Often the existing research has been carried out with a politically motivated background—one in favor of accepting a human right to water too easily. Although this study also concludes with a positive presumption in favor of a human right to water, it does so in a more nuanced way, attempting to keep political considerations out of the legal analysis. Often the existing research has also been too narrowly focused—either mainly concentrating on international rather than domestic law; or failing to tackle the issue from more than one discipline. Existing literature often focuses either on the legal status of a human right to water, assesses the economic prospects of water privatization or makes a philosophical claim in favor of or against global water solidarity. This book combines these different disciplines and techniques in order to find holistic answers: it deploys legal analysis, economic reasoning and philosophical argumentation in combination. Only

---

[17] For the introduction of the mandate, see HRC, *Human rights and access to safe drinking water and sanitation*, A/HRC/RES/7/22, adopted on 28 March 2008. For an overview of the mandate's work, also see the webpage of the Office of the High Commissioner for Human Rights (OHCHR), http://www2.ohchr.org/english/issues/water/iexpert/index.htm.

[18] See the curriculum vitae of Mrs. de Albuquerque at the OHCHR webpage, http://www.ohchr.org/en/Issues/WaterAndSanitation/SRWater/Pages/CatarinaDeAlbuquerque.aspx.

holistic answers—answers drawing on insights across legal regimes and different disciplines—will suffice in tackling one of the biggest challenges faced by modern humankind.

Finally, the book takes into account the recent developments that the right has experienced during the calendar years of 2010 and 2011: the recent UNGA Resolution[19] and the Resolution of the HRC[20] as well as the renewal of the Independent Expert's mandate (now called: Special Rapporteur).[21] Have these developments given the right finally the status and meaning that many have claimed for it since a long time?

References

Barlow M, Clarke T (2002) Blue Gold: the fight to stop the corporate theft of the world's water. New Press, New York

Department for International Development (DFID) (2001) Addressing the water crisis: healthier and more productive lives for poor people. http://webarchive.nationalarchives.gov.uk/+/http://www.dfid.gov.uk/pubs/files/tspwater.pdf

Dugger, CW (2008) Cholera epidemic sweeping across crumbling Zimbabwe. http://www.nytimes.com/2008/12/12/world/africa/12cholera.html?_r=0

Kessler S (2007) Wer hat Teil am Blauen Gold? Bemerkungen zur menschenrechtlichen Dimension des Rechts auf Wasser? In: Rudolf B (ed) Menschenrecht Wasser? Peter Lang, Frankfurt am Main, pp 185–194

Langford M (2006) Expectation of plenty: response to Stephen Tully. Neth Q Hum Rights 24 (3):473–479

McNeil DG Jr (2010) Cholera outbreak kills 150 in Haiti. http://www.nytimes.com/2010/10/23/world/americas/23cholera.html

Petrova V (2006) At the frontiers of the rush for Blue Gold: water privatization and the human right to water. Brooklyn J Int Law 31(2):577–613

Rudolf B (2007) Menschenrecht Wasser: Herleitung, Inhalt, Bedeutung, Probleme. In: Rudolf B (ed) Menschenrecht Wasser? Peter Lang, Frankfurt am Main, pp 15–43

Sontag D (2012) In Haiti, global failures on a cholera epidemic. http://www.nytimes.com/2012/04/01/world/americas/haitis-cholera-outraced-the-experts-and-tainted-the-un.html

United Nations (UN) (2009) The Millennium Development Goals Report 2009. United Nations, New York. http://www.un.org/millenniumgoals/pdf/MDG_Report_2009_ENG.pdf

United Nations (UN) (2012) The Millennium Development Goals Report 2012. United Nations, New York. http://www.un.org/millenniumgoals/pdf/MDG%20Report%202012.pdf

United Nations Development Programme (UNDP) (2006) Human Development Report 2006: beyond scarcity – power, poverty and the global water crisis. http://hdr.undp.org/en/media/HDR06-complete.pdf

United Nations Children's Fund (UNICEF), World Health Organization (WHO) (2012) Progress on drinking water and sanitation: 2012 update. http://www.unicef.org/media/files/JMPreport2012.pdf

World Health Organization (WHO) (2012) Glass 2012 Report: UN-water global analysis and assessment of sanitation and drinking water – the challenge of extending and sustaining services. http://whqlibdoc.who.int/publications/2012/9789241503365_eng.pdf

[19] UNGA, *The human right to water and sanitation*, A/RES/64/292, 3 August 2010.

[20] HRC, *Human rights and access to safe drinking water and sanitation*, A/HRC/RES/15/9, 6 October 2010.

[21] HRC, *The human right to safe drinking water and sanitation*, A/HRC/RES/16/2, 8 April 2011.

Chapter 2
The Current Legal Status of the Right to Water

2.1 The Right to Water in Europe

The starting point of this legal analysis is the legal order which is the most proximate and maybe the most developed one: the European legal framework. Beginning with examples of the status of the right to water in three selected Member States of the European Union (EU), the study then turns to the legal protection offered to the right to water on the European level itself, namely by the EU (European Law in a narrow sense) and in the context of the European Convention for the Protection of Human Rights and Fundamental Freedoms (ECHR) and the European Social Charter (ESC) (European Law in a broader sense).

2.1.1 The Right to Water in National Law of European States

The orders of the European Member States are often said to be, all-in-all, fairly homogenous compared to the rest of the world. This book scrutinizes whether this statement applies as well to the right to water with its strong socio-economic dimensions. Socio-economic rights are more contested in their legal status compared to civil and political rights,[1] sometimes even in their nature as rights at all.[2] The selection of the three country studies is motivated by the attempt to analyze three very different approaches to a subjective right to water within Europe. As the study will show, the three selected legal systems—Germany, Belgium and France—constitute three distinctly different approaches to give a subjective right to water legal effect.

*All cited webpages were last accessed on 1 August 2013.

[1] Goldsmith (2001), p. 1212; Mehta (2005), p. 238.

[2] Cranston (1973), pp. 19 ff.; also see *id.* (1967), p. 51; Dennis and Stewart (2004), p. 465.

P. Thielbörger, *The Right(s) to Water*,
DOI 10.1007/978-3-642-33908-0_2, © Springer-Verlag Berlin Heidelberg 2014

2.1.1.1 The Right to Water in Germany

The general situation of water access and water supply in the Federal Republic of Germany is not very problematic. There is no water shortage in Germany, especially when compared with some other (southern) European countries.[3] According to official numbers assessed by the Federal government, in 2010, more than 99 % of the German population was connected to the water supply system.[4] While water prices in Germany vary significantly throughout Germany,[5] they are generally amongst the highest in Europe.[6] The micro-biological quality of water in Germany is generally considered to be of an excellent standard.[7] The administrative part of the water supply is generally a task of the municipalities ("örtliche Gemeinschaft") in the context of the guarantee of municipal self-administration ("kommunale Selbstverwaltungsgarantie"),[8] as laid down in Art. 28 par. 2 of the German constitution ("Grundgesetz", GG).

The Protection of the Right to Water through German Laws

The German constitution protects a variety of basic and human rights ("Grundrechte") in its articles 1–19. However, there is no direct reference to an explicit subjective right to water or a specific entitlement to a basic minimum water supply within these articles, nor elsewhere in the constitution. Silke Ruth Laskowski has recently forcefully argued that the right to life under the German constitution (Art. 2 par. 2)[9] should be considered stronger as the derivative basis for the human right to water in the German context; such an entitlement would be limited, however, to the basic amount necessary for human existence.[10]

---

[3] See BMU (2011), p. 4; Köck (2012), p. 140.

[4] Federal Statistical Office, Wasserwirtschaft: Öffentliche Wasserversorgung in Deutschland von 1991 bis 2010, 31 January 2013, https://www.destatis.de/DE/ZahlenFakten/GesamtwirtschaftUmwelt/Umwelt/UmweltstatistischeErhebungen/Wasserwirtschaft/Tabellen/Wasserabgabe1991_2010.html;jsessionid=A2DB2E307B2247AF4BD595D9D9F386EE.cae3.

[5] Dreyer and Bartl (2010), p. 2553; Busse von Colbe (2011), p. 575.

[6] Busse von Colbe (2011), p. 575.

[7] In some areas of Germany, groundwater can even be consumed without previous chemical treatment; see for instance for the case of Bavaria, Emmerich-Fritsche (2007), pp. 1–2.

[8] Jarass and Pieroth (2012), Art. 28 par. 13a; Henneke (2011), Art. 28 par. 72.

[9] Art. 2 par. 2 Constitution of the Federal Republic of Germany, http://www.gesetze-im-internet.de/englisch_gg/englisch_gg.html: "Every person shall have the right to life and physical integrity. Freedom of the person shall be inviolable. These rights may be interfered with only pursuant to a law."

[10] Laskowski (2010), p. 905.

Specific Provisions on Water in Ordinary Law

More concretely, there are many German ordinary laws dealing with the topic of water. As the legislative competence in the sector of water law is a split competence between the German national State and the 16 German federal States ("Länder"), there are some relevant German national laws, being completed by water laws on the level of the 16 German federal States, which implement many of the guidelines of the EU Water Framework Directive.[11]

The most important one of these federal laws on water is the Federal Water Act (or: Federal Hydrological Act) ("Wasserhaushaltsgesetz", WHG).[12] The act regulates the use and protection of surface water, groundwater and coastal water. Section 8 of the Act emphasizes that any use of water as defined in the act requires permission by a German administrative body. Section 12 par. 1 WHG states the conditions under which permission must be denied, in particular if the use would create lasting damage to the water or is otherwise in violation of public law. As the highest German administrative court ("Bundesverwaltungsgericht") stated, such permission must be denied, if the intended use of water constitutes a threat to the public welfare, particularly to the public water supply, which includes any act potentially dangerous for public health.[13] It is interesting from a human rights perspective to note that there is no entitlement to receive permission for a specific use, even if the criteria for exclusion of section 12 par. 1 WHG are not given. Even in this case, to give permission is to a certain discretion of the authorities (section 12 par. 2 WHG; so called "Bewirtschaftungsermessen", which is in essence a form of repressive prohibition with the possibility for exceptions).[14] The exercise of this discretion is to be guided by the principles laid down in section 6 WHG, such as precaution and sustainability, and must serve the aim to preserve water as a common good.[15] Only one use is exempted from the need of permission, namely the use of owners to use water for their own domestic purposes ("Eigentümergebrauch"), if the water quality remains unaffected (section 26 WHG).

German criminal law also deals with the issue of water explicitly. The Criminal Code ("Strafgesetzbuch") states in section 324 par. 1 that whoever pollutes a body of water or otherwise detrimentally alters its quality without authorization will be punished with imprisonment of up to 5 years or with a fine. Even attempt and

---

[11] See the EU's Water Framework Directive, more specifically European Parliament and Council of the European Union, *Directive 2000/60/EC of the European Parliament and of the Council of 23 October 2000 establishing a framework for Community action in the field of water policy*, OJ 2000 L 327, 1, establishing a framework for Community action in the field of water policy.

[12] Wasserhaushaltsgesetz (WHG), BGBl. I, 1110, 1386, 27 July 1957; see the latest version in BGBl. I, 2449, 5 December 2012; see on this more comprehensively Knopp (2010).

[13] BVerwG, Case No. 4 C 30/88, 17 March 1989, BVerwGE 81, 347, 350.

[14] Czychowski and Reinhardt (2010), section 8 par. 3. For a fundamental critique of this regime which gives the administration much discretion, see Papier (2012), Art. 14 par. 430.

[15] Czychowski and Reinhardt (2010), section 6 par. 2 and 26.

negligence are punishable.[16] In severe cases, as in section 330 of the Criminal Code, the punishment can be up to 10 years, and, if the perpetrator endangers the public water supply, a serious case can regularly be assumed.[17]

However, the German water-related laws do not address the question of a *subjective* right to water, thus a right for the individual. They deal more with the general planning, cultivation and government control of water-bodies and watercourses, with standard values of drinking water quality,[18] implementing European demands laid down, for instance, in the European Drinking Water Directive,[19] or with the criminal treatment of water contaminators. In order to spot legal approaches to a subjective right to water, one has to look into more general legal concepts under German law.

Related Legal Concepts: "Daseinsvorsorge" and "Sozialstaatsprinzip"

One of these concepts is the more than 70 years old concept of "Daseinsvorsorge", which goes back to Ernst Forsthoff's idea of "Daseinsfürsorge".[20] It is most important under the German constitutional order until today.[21] The closest English translation is "services of general (economic) interest"; sometimes also the term "services of elementary requirements" is used. It covers a variety of fields, such as energy supply, public transports, cemeteries or cultural and sports facilities.

Prima facie the term "Daseinsvorsorge" can be understood in a purely descriptive way,[22] namely as that part of a State's welfare which provides the individual with very basic goods and services. However, in contrast to the French idea of the "service public",[23] the concept of "Daseinsvorsorge" in Germany has sometimes been seen more as a sociological than as a juristic term.[24] One reason for this view is

[16] § 324 German Criminal Code ("Water pollution"), http://www.gesetze-im-internet.de/englisch_stgb/englisch_stgb.html#p2732: "(1) Whosoever unlawfully pollutes a body of water or otherwise alters its qualities in a negative manner shall be liable to imprisonment not exceeding 5 years or a fine. (2) The attempt shall be punishable. (3) If the offender acts negligently the penalty shall be imprisonment not exceeding 3 years or a fine."

[17] § 330 German Criminal Code ("Aggreved cases of environmental offence"), http://www.gesetze-im-internet.de/englisch_stgb/englisch_stgb.html#p2732: "In especially serious cases of an intentional offence under sections 324 to 329 the penalty shall be imprisonment from 6 months to 10 years. An especially serious case typically occurs if the offender [. . .] 2. endangers the public water supply".

[18] See in this regard especially the German Regulation on Drinking Water ("Trinkwasserverordnung"), 21 May 2001, BGBl. I (2001), p. 959.

[19] See Council of the European Union, *Council Directive 98/83/EC of 3 November 1998 on the quality of water intended for human consumption* (European Drinking Water Directive), OJ 1998 L 330, 32.

[20] See Forsthoff (1938); *id.* (1959).

[21] Püttner (2002), p. 3; Kersten (2005), p. 544.

[22] Rüfner (2006), par. 10.

[23] The concept lies outside the focus of this analysis. For more detailed information, see Le Nestour and Zinow (1994); Bullinger (2003).

[24] Ruge (2003).

that the German legislator has been reluctant to incorporate the term "Daseinsvorsorge" into concrete German laws,[25] although the German Constitutional Court ("Bundesverfassungsgericht") emphasized that "Daseinsvorsorge" is an issue of constitutional importance.[26]

Ernst Forsthoff developed the term in his time in order to submit even purely distributive administrative acts and positive State action of each kind to the regime of public law. Administrative action should legally be treated as administrative action, even if the State appeared in a purely private guise.[27] Administrative action in this sense refers to the State taking care of the life possibilities of citizens by distributive means.[28] Traditionally, "Daseinsvorsorge" is understood to include an element of the citizens' needs for certain goods that the individual cannot achieve by his own means, due to an increasingly industrialized environment.[29] In this way, the market economy system creates some unintentional effects, and the lack of sufficient supply in some areas must be counteracted by the "caring hand" of the welfare State.[30] The State has to create structures and appropriate frameworks for those areas in which the free market alone cannot effectively guarantee the supply of essential goods and services.

The most relevant question for this examination is to what extent subjective rights are included in or can be derived from the concept of "Daseinsvorsorge". It is broadly assumed that Forsthoff's conception was meant to be used as apposite or contrary to an economic constitution which is based on subjective economic liberty rights.[31] In this sense the idea of "Daseinsvorsorge" must more be understood as an attempt to socially discipline the State.[32] The focus of this concept lies in the State's own responsibility, not on the rights or entitlements of the individual. As a counterpart to certain benefits, the State has to take certain duties on board that it can only fulfil because of the privileges that are also granted to it. On its own, however, the concept of "Daseinvorsorge" does not grant individual entitlements.

Secondly, many municipal laws order a compulsory connection and usage system ("Anschluss- und Benutzungszwang").[33] If needed, the municipalities can determine that all real estate must be connected to water supply and similar

---

[25] *Ibid.*, p. 354.

[26] BVerfG, Case No. 1 BvL 28/82, 20 March 1984, BVerfGE 66, 248, 258.

[27] Forsthoff (1959), p. 10.

[28] Wolff et al. (2007), p. 55.

[29] Forsthoff (1959), p. 38.

[30] Ambrosius (2000), p. 15.

[31] Mestmäcker (1998), pp. 638 and 648; Bullinger (2003), p. 598, with further references; see also Ringwald (2008), pp. 30–70, with a detailed analysis of Forsthoff's conception of "Daseinsvorsorge".

[32] See Kersten (2005), pp. 543 ff.

[33] See for instance § 4 Berliner Betriebe-Gesetz (BerlBG), http://www.bvg.de/index.php/de/binaries/asset/download/21576/file/1-1; § 11 Gemeindeordnung für Baden-Württemberg (GemO BW), http://www.landesrecht-bw.de/jportal/?Quelle=jlink&query=GemO+BW&psml=bsbawueprod.psml&max=true&aiz=true. See for more details Musil and Kirchner (2012), par. 505–510; Seewald (2006), pp. 66 ff.

infrastructure. As a typical consequence, construction projects can only be permitted if the water supply is ensured, usually by connection to the local water distribution system. This concept, which aims at increasing water supply coverage, however, is mainly designed as a duty (for building owners) rather than as a subjective right to water for all citizens.[34] If water supply is operated through the private sector, there is also a mirror obligation for the private sector: companies providing vital services, such as water, have an obligation to contract with customers seeking their service ("Kontrahierungszwang").[35]

The third concept that is relevant for the question of a subjective right to water within the German legal framework is the principle of the social State enshrined in Art. 20 of the constitution ("Sozialstaatsprinzip"). Despite only being mentioned briefly in the text of the GG,[36] it is accepted as one of the founding principles of the German constitution.[37] In interpreting this provision, the norms of the European Social Charter must be considered.[38] The indeterminate nature of this provision means that there is an even more pressing need for it to be concretized by the legislator, the executive and the judiciary than other constitutional principles.[39]

Generally, the principle of the social State does not create concrete entitlements or rights to social benefits of a certain amount by itself.[40] Even combined with certain fundamental rights as laid down in the GG, the principle of the social State creates positive rights towards certain goods or services only in exceptional cases,[41] which have been elaborated by the German judiciary, which will be examined in the following section. The principle of the social State on its own, however, is better understood as one of the State's guidelines for its political and legal action rather than as a basis for concrete entitlements for citizens.

---

[34] Admittedly, this is also reflection of the protection of a right, for instance, the right of future tenants.

[35] See for this institute comprehensively Busche (1999); more specifically for the case of water, see Brinkmeier (2002).

[36] See Art. 20 par. 1 Constitution of the Federal Republic of Germany, http://www.gesetze-im-internet.de/englisch_gg/englisch_gg.html: "The Federal Republic of Germany is a democratic and social federal State"; and Art. 28 par. 1 s. 1: "The constitutional order in the Länder must conform to the principles of a republican, democratic and social State governed by the rule of law, within the meaning of this Basic Law."

[37] Jarass and Pieroth (2012), Art. 20 par. 112; Herzog and Grzeszick (2012), Art. 20 par. 2.

[38] BVerfG, Case No. 1 BvR 1213/85, 2 March 1993, BVerfGE 88, 103, 112; Jarass and Pieroth (2012), Art. 20 par. 112.

[39] BVerfG, Case No. 2 BvR 458/80 and 2 BvR 486/80, 19 October 1983, BVerfGE 65, 182, 193; BVerfG, Case No. 1 BvL 2/82, 22 October 1985, BVerfGE 71, 66, 80.

[40] BVerfG, Case No. 2 BvL 5/00, 8 June 2004, BVerfGE 110, 412, 455; Jarass and Pieroth (2012), Art. 20 par. 124.

[41] BVerfG, Case No. 1 BvR 624/56, 3 December 1969, BVerfGE 27, 253, 283; BVerfG, Case No. 1 BvL 20/84, 1 BvL 26/84 and 1 BvL 4/86, 29 May 1990, BVerfGE 82, 60, 80; Herzog and Grzeszick (2012), Art. 20 par. 28.

The Protection of the Right to Water through German Courts

One of the cases to specify the social State principle developed by the judiciary is its combination with Art. 1 par. 1 GG,[42] the German provision on human dignity. Situated at the very beginning of the German constitution, it recognizes the paramount importance of fundamental rights and the constitution's humanistic orientation.

Human dignity and the social State principle read together constitute an obligation to guarantee the "very basic requirements for an existence in human dignity" ("Existenzminimum"). After initial scepticism relating to the objective nature of the social State principle,[43] this "Existenzminimum" is in the meantime recognized in the decisions of the German Constitutional Court.[44] The Federal Administrative Court also acknowledges the existence of a constitutional entitlement to a minimum of benefits.[45] This minimum level is not limited to the level necessary for bare survival, but encompasses the so called "socio-cultural minimum existence level",[46] which enables a life in dignity, including participation in cultural life. However, the "Bundesverfassungsgericht" has so far only dealt with questions of financial aid to ensure a minimum existence and has made the lack of financial means a prerequisite for the realization of the "Existenzminimum".[47]

In other cases, the courts had to deal more specifically with the question of water. In an early ruling of the federal republic, the constitutional court already accepted in its famous "Nassauskiesung"-decision that the supply of drinking water of sufficient quantity and quality is essential for the life of the population in general,[48] and that water is one of the most important fundaments of all human, animal and vegetable life.[49] Therefore, the court accepted the State's decision to establish a separate regime for groundwater without giving particular rights to the landowners.[50] It regarded groundwater, being the basis for a sufficient water supply,

[42] Art. 1 par. 1 Constitution of the Federal Republic of Germany, http://www.gesetze-im-internet.de/englisch_gg/englisch_gg.html: "Human dignity shall be inviolable. To respect and protect it shall be the duty of all State authority."

[43] BVerfG, Case No. 1 BvR 220/51, 19 December 1951, BVerfGE 1, 97, 104 stated in the early days of the constitution (1951) that the social principle on its own can only create entitlements, if the State arbitrarily violates this obligation; also see BVerfG, Case No. 1 BvR 762/85, 20 May 1987, BVerfGE 75, 348, 360.

[44] See, on the "Existenzminimum" of asylum-seekers, most relevantly BVerfG, Case No. 1 BvL 20/84, 1 BvL 26/84 and 1 BvL 4/86, 29 May 1990, BVerfGE 82, 60, 80; and most recently *id.*, Case No. 1 BvL 10/10 and 1 BvL 2/11, 18 July 2012, in: *Neue Zeitschrift Verwaltungsrecht* 31(16):1024–1033 (with comment by Tiedemann); support for this approach by Herdegen (2012), Art. 1 par. 1 and 121.

[45] BVerwG, Case No. V C 78/54, 24 June 1954, BVerwGE 1, 159, 161.

[46] BVerfG, Case No. 1 BvL 18/93, 1 BvL 5/94, 1 BvL 6/94, 1 BvL 7/94, 1 BvR 403/94 and 1 BvR 569/94, 24 January 1995, BVerfGE 92, 112, 114.

[47] BVerfG, Case No. 1 BvL 20/84, 1 BvL 26/84 and 1 BvL 4/86, 29 May 1990, BVerfGE 82, 60, 80; Jarass and Pieroth (2012), Art. 20 par. 124.

[48] BVerfG, Case No. 1 BvL 77/78, 15 July 1981, BVerfGE 58, 300, 344. See also Baur (1982); Knauber (1984).

[49] BVerfG, Case No. 1 BvL 77/78, 15 July 1981, BVerfGE 58, 300, 341.

[50] *Ibid.*, 301.

as a common good with particular value for the population as a whole.[51] Essentially the court affirmed the legislative decision to exclude groundwater from the scope of application of the right to property. It also confirmed that excluding groundwater from property does not constitute an expropriation which would have required compensation for landowners. A few years later, the court accepted in the case of "Grosser Erftsverband" limitations of the rights of mining companies in favour of water resource management as a consequence of Art. 14 par. 2 of the German constitution, which demands that property entails obligations, and in overall terms has to serve the public good.[52]

This opinion was confirmed by the court again in the 1990s in its "Wasserpfennig"-decision,[53] where the court approved a special tax on the use of groundwater by private persons, because water was a scarce natural resource that belonged to the population as a whole.[54] Altogether, the court stated that drinking water supply is dependent on high quality groundwater, and that this water should thus be used to gain drinking water, and approved of higher prices for those who extract groundwater for economic or industrial means.

The "Bundesverwaltungsgericht" has followed this opinion. It emphasized the outstanding importance of water as a common good[55] and regarded the supply of basic water services as a vitally important element of the "Daseinsvorsorge" for big parts of the population. According to the court, the preservation and protection of water was a basic task of the federal State.[56]

Analysis of the Protection of the Right in Germany

The German legal system does not offer an explicit protection of the right to water. It largely operates with and relies on traditional legal concepts, mainly the principles of "Daseinsvorsorge" and "Sozialstaatsprinzip", which in combination with some general fundamental rights can give rise to concrete subjective rights.

This might be a surprising outcome at first sight. The German constitution, being developed under the impression of the previous Nazi-regime, places great emphasis on individual fundamental rights; so why did it not consider a right to access water? This omission is less surprising, if one considers that the German legislator did not feel the need to establish a more concretely shaped right to water, owing to the fact that Germany does not face water scarcity. However, this approach leaves many powers and freedoms to the judiciary, which has to give vague terms concrete legal content. This is not unproblematic, as the three important German High Courts, the

[51] *Ibid.*, 339.

[52] BVerfG, Case No. 1 BvR 394/58, 29 July 1959, BVerfGE 10, 89.

[53] BVerfG, Case No. 2 BvR 413/88 and 2 BvR 1300/93, 7 November 1995, BVerfGE 93, 319.

[54] BVerfGE 319, 350.

[55] BVerwG, Case No. 6 C 2/97, 17 December 1997, BVerwGE 106, 64.

[56] *Ibid.*

"Bundesverfassungsgericht", the "Bundesverwaltungsgericht" and the "Bundesgerichtshof" (BGH), Germany's highest civil court, have unanimously seen the concretization of general constitutional principles such as the principle of the social State generally as a binding task for the legislator.[57] This should even more apply in cases where the concretization of these principles involves considerable financial impacts for the public authorities, which in the field of water supply is certainly the case.

This "minimalist" legal approach does of course not mean that the German legislator in any way intends to deny factual access to essential basic services for its population. It should more be understood as a consequence of a perceived lack of necessity, due to the adequate levels of water supply available in Germany. By contrast, on the international level—where the German government senses more of an urgent need for change—it has taken a very proactive role. Since 2006, Germany took several joint actions together with the Spanish government for the recognition of a human right to water on the international level and took the initiative that ultimately lead to the establishment of the mandate for an Independent Expert on the issue of human rights obligations related to access to safe drinking water and sanitation.[58] However, for a more coherent and credible approach Germany should join those European countries that have formally recognized a right to water in their national legislation.[59] This is the logical domestic counterpart of Germany's engagement for the recognition of a human right to water on the international level. Its political meaning should not be underestimated.

Given the well-designed web that the German courts have created around the general principles of "Daseinsvorsorge" and the principle of the "Sozialstaat", such an introduction would not entail significant legal change. However, it would contribute to legal certainty for citizens, who are often not aware of the abstract principles that the judiciary has developed. In case of an inclusion into the constitution, and not only into ordinary laws, the legislator could also emphasize that the right to water has the rank of a constitutional value. However, such a constitutional

---

[57] BVerfG, Case No. 1 BvL 30/76, 3 April 1979, BVerfGE 51, 115, 125; BVerwG, Case No. 1 C 136/80, 24 June 1982, BVerwGE 66, 29, 32; BGH, Case No. VI ZR 349/88, 19 September 1989, BGHZ 108, 305, 310.

[58] See, from the perspective of the German Federal Foreign Office, AA (2010), p. 19, and the article "Water—a human right" on its website, http://www.auswaertiges-amt.de/EN/Aussenpolitik/Menschenrechte/MRVN-Wasser_node.html.

[59] In Europe, the right to water is formally recognized in Belgium, Finland, France, Russia, Spain, Sweden, Ukraine and United Kingdom (UK). Social tariffs for water exist in Austria, Bulgaria, Greece, Hungary, Luxembourg, Malta, Portugal, The Netherlands and in United Kingdom; see Report of the World Water Council "Right to Water—Moving towards a global consensus", http://www.worldwaterforum5.org/fileadmin/WWF5/Documents/Story_RTW_CD_March07.pdf. See also OHCHR, *Study on human rights obligations related to equitable access to safe drinking water and sanitations—Contributions received*, http://www2.ohchr.org/english/issues/water/contributions.htm; and the overview of the World Water Council, *Right to Water—Moving towards a global consensus*, March 2007, http://www.worldwaterforum5.org/fileadmin/WWF5/Documents/Story_RTW_CD_March07.pdf.

amendment needs a two third majority according to Art. 79 par. 2 of the German constitution and is thus not easy to achieve. However, it is not clear that any political party in Germany would be generally opposed to such an approach.[60] The inclusion of the right to water in the German constitution could become important, if different values and different rights clash with each other, whereby constitutional values would potentially prevail over values that are protected only by ordinary laws, or where national provisions clash with international ones.[61] The general reluctance of the German legislator to expand the canon of fundamental rights too much is understandable. However, the human need for water is more essential than almost any other. As a consequence, this objection against constitutional inflation does not apply in the particular case of drinking water.

On the level of ordinary law, however, the German legislator and judiciary have concretized the most crucial elements of the normative content of the right to water[62]: Sufficient and continuous *availability* of water access for everyone is generally ensured through the concept of "Daseinsvorsorge". The high *quality* of water is upheld through the requirement of permission, laid down in the WHG, for uses that could be in conflict with the public interest, and the penalization of water contamination in the Criminal Code. Finally, the *physical accessibility* is largely ensured through the "Anschluss- und Benutzungszwang": not only are all public buildings connected to water, but connecting to the local water supply system is also a prerequisite for building new private households. In case of privatization, private suppliers must contract with consumers due to the "Kontrahierungszwang". The *economic accessibility* finally is ensured through the judiciary by referring to the social State principle and Art. 1 of the constitution: the State must ensure the socio-economic existential minimum for all people.

Thus, it is much regrettable that an outstanding recognition of a right to water in the German constitution is missing, in particular as Germany decided in the 1990s to include environmental protection in the constitution.[63] This was a missed chance to also include an explicit right to water. However, Germany is an example of how, even without an accentuated recognition, but through ordinary laws and an efficient judiciary, the right can receive effective and comprehensive legal protection.

---

[60] See the most recent press statement of the green political party in Germany, Bündnis 90/Die Grünen, "Wasser ist ein Menschenrecht! Keine Nötigung zur Wasserprivatisierung", 16 February 2013, http://www.gruene-nrw.de/details/nachricht/wasser-ist-menschenrecht-keine-noetigung-zur-wasserprivatisierung.html. See also the motion of the commission for development of the German Bundestag "Wasser und Ernäherung sichern" ("Assuring Water and Food"), 27 March 2012, Drucksache (Official Records) 17/9153.

[61] See Sect. 3.4.4 of this book.

[62] See for an in-depth discussion of the normative content, Sect. 2.3.2.2 of this book.

[63] Introduced in 1994, Art. 20a Constitution of the Federal Republic of Germany, http://www.gesetze-im-internet.de/englisch_gg/englisch_gg.html reads now: "Mindful also of its responsibility toward future generations, the State shall protect the natural foundations of life and animals by legislation and, in accordance with law and justice, by executive and judicial action, all within the framework of the constitutional order." For legal consequences, see for instance Bernsdorff (1997); Gassner (2011).

2.1.1.2 The Right to Water in Belgium

A second remarkable example within the European legal orders is that of the Kingdom of Belgium. Although Belgium is generally considered to fully comply with human rights obligations related to drinking water,[64] Belgium is an extraordinarily interesting case study, mainly for two reasons. Firstly, Belgian courts have made themselves a name over the last years in protecting a right to water and in being prepared to implement it, even where the law is ambiguous. Secondly, in 2005, the federal Belgian parliament adopted the so called "Water Resolution", which recognizes access to safe drinking water as a human right that should be included in the constitutional text[65]: "le droit à l'accès à l'eau potable en quantité et d'une qualité suffisante en tant que droit fondamental de l'être humain". Such a constitutional amendment would, if finally passed, make Belgium the first European Union country to include the right to water explicitly in its constitution. However, even 8 years later, an agreement for such a constitutional amendment has still not been reached.

The Protection of the Right to Water through Belgian Law

Beginning in the early 2000s, the Belgian public has been instrumental in the gradual strengthening of the right to water under Belgian law. In 2001 and 2002, a "Water Manifesto" ("Manifeste de l'eau") was set up, asking for water to be considered as part of the common heritage of mankind, including a guaranteed access to water for present and future generations. 120,000 citizens subscribed to this manifesto, as did the governments and parliaments of the three Belgian regions.[66] The summit of this pro-"right to water" movement is certainly the mentioned adoption of the Water Resolution intending to make an amendment of Art. 23 of the Belgian constitution with regard to water. As long as this amendment is still on its way,[67] Art. 23 of the Belgian constitution in its previous version remains nevertheless the most important norm with respect to water. Art. 23 of the constitution states that all citizens should be able to live in a dignified manner and therefore have the right to the protection of a safe environment and health.[68] The

[64] See for instance OECD (2007), p. 170.

[65] See Chamber of Representatives of Belgium (Commission spéciale mondialisation), *Proposition de résolution: Accès à l'eau pour chacun,* Doc. 51 1666/002, 14 March 2005; also see Barlow (2007), p. 174.

[66] Armeni (2008), p. 5; OHCHR (2007).

[67] See the status of the amendment and the last draft of 18 November 2011 at http://www.senate.be/www/?MIval=/dossier&LEG=5&NR=1336&LANG=fr. Given the long timeframe, one might by now seriously doubt that the constitutional amendment will ever be decided.

[68] Art. 23 Constitution of the Kingdom of Belgium, 2012, http://www.dekamer.be/kvvcr/pdf_sections/publications/constitution/grondwetEN.pdf: "Everyone has the right to lead a life in keeping with human dignity. To this end, the laws, federate laws and rules referred to in Article 134

competence to implement this right is given to both the federal State government and the three Belgian regions Wallonia, Flanders and Brussels.[69] All three regions have passed laws to implement this legal demand. However, they have chosen very different means and placed different emphasis on the right's normative content. The following section will not conduct a complete analysis of the three different Belgian water regimes, but highlight specific elements of the regimes that are distinct for the different approaches.

In *Flanders*, legislation recognizes the right to a minimal supply of natural gas, electricity and also water for everybody. In the case of water this means that every person is entitled to receive a minimum amount of drinking water *free of charge* per year (15 m^3).[70] Allotting water to individuals rather than households increases equality in the allocation of free water[71] and ultimately favors families with many children.[72] Flanders has also adopted a progressive water-pricing mechanism.[73] The tariff depends on the amount consumed: a basic fee covers the fixed costs of connection, no fees are charged for the first 15 m^3 of consumption, and the cost for additional consumption is designed as a variable cost, depending on how big this additional consumption is. These composed variable costs are of different blocks of water prices, determined by the distribution companies and the regional authorities.[74] Flanders has also launched projects on rainwater storage in order to raise awareness of water scarcity.[75]

In *Wallonia*, the "*code de l'eau*" states that each person is entitled to drinking water of quality and in sufficient quantity for their nutrition, their household needs and their health.[76] There is no free minimum amount as in Flanders, but progressive

guarantee economic, social and cultural rights, taking into account corresponding obligations, and determine the conditions for exercising them. These rights include among others: (1) the right to employment and to the free choice of an occupation within the context of a general employment policy, aimed among others at ensuring a level of employment that is as stable and high as possible, the right to fair terms of employment and to fair remuneration, as well as the right to information, consultation and collective negotiation; (2) the right to social security, to health care and to social, medical and legal aid; (3) the right to decent accommodation; (4) the right to the protection of a healthy environment; (5) the right to cultural and social fulfillment."

[69] *Ibid.* ("laws, federate law and rules referred to in Article 134"); Ouvry (2007), p. 1.

[70] See Art. 3.4 Décret de la Communauté Flamande concernant diverses mesures d'accompagnement du budget 1997, 20 December 1996, in: *Moniteur belge*, 31 December 1996, 3rd edition, quoted after Armeni (2008), p. 1; also see Smets (2006b), p. 86; *id.* (2008a), p. 221.

[71] See Armeni (2008), p. 2; Smets (2005), p. 31.

[72] Smets (2008a), p. 90.

[73] *Ibid.*, p. 114.

[74] Armeni (2008), p. 2; see also OECD (2006), p. 115.

[75] See for instance Cornut et al. (2006).

[76] Art. 1.1 Décret relatif au Livre II du Code de l'Environnement constituant le Code de l'Eau, in: *Moniteur belge*, 23 July 2004. See similar wording also in Art. 1.2 Décret de la Région Wallonne relatif au cycle de l'eau et instituant une société publique de gestion de l'eau, in: *Moniteur belge*, 22 June 1999.

water pricing is used for water supply and sanitation, with a first amount of 30 m^3 per household per year at a lower rate.[77] This approach essentially subsidizes small users at the cost of large (often industrial) users.[78] Prices must reflect the "true costs" of water including the cost of water production, water distribution and water preservation.[79] Furthermore, a "Social Fund for Water" provides financial support for those households having difficulties to pay their water bills.[80] This fund is financed by a small tax per m^3 of billed public water supply.[81] However, in 2004, only a few households were eligible for support from the fund.[82] Since then, the numbers of beneficiaries and the average amount of received aid have increased.[83] In 2010, the total amount of aid accrued to 1.93 million EURO.[84] A problem with respect to non-discrimination is, however, that the fund in its beginnings was only available to the francophone area of the region, not to the German-speaking community.[85]

The Walloon engagement for water has also entered the international scene: The Walloon Minister of Environment announced during the 4th World Water Forum in Mexico in 2006 that the regional government was planning to introduce another tax of 1.25 EURO cents per m^3, which would be used to provide assistance for water projects in selected developing countries.[86] In 2008, this proposal was realized by establishing a "fonds de solidarité international pour l'eau".[87]

In the last region, the Brussels Capital Region, the legislation has also recognized the right to the distribution of drinking water of an acceptable quality for household consumption.[88] A progressive water pricing system is based on three

---

[77] *Ibid.*; Art. D.228 Code de l'eau, http://environnement.wallonie.be/legis/Codeenvironnement/codeeaucoordonne.htm; Smets (2008a), p. 94.

[78] OHCHR (2007), p. 3.

[79] Armeni (2008), p. 3.

[80] Art. D.234 Code de l'eau, http://environnement.wallonie.be/legis/Codeenvironnement/codeeaucoordonne.htm; also see Décret relative á la création d'un Fond Social de l'Eau en Région Wallonne, 20 February 2003, in: *Moniteur belge*, 19 March 2003, also available at http://www.environnement.wallonie.be/legis/eau/eafon002.htm.

[81] Art. D.239 Code de l'eau, http://environnement.wallonie.be/legis/Codeenvironnement/codeeaucoordonne.htm; also see Smets (2006b), p. 86; *id.* (2008a), p. 111.

[82] Ouvry (2007), p. 2; Smets (2006b), p. 85.

[83] Boury (2008), p. 7.

[84] See Aquawal (2012), p. 40. It should be noted, however, that the maximum amount of financial aid per outstanding water bill was increased in 2008. Since then, the number of interventions dropped (from 11,008 in 2009 to 8,312 in 2010), but the average amount of financial aid granted increased (from 175 EURO in 2009 to 232 EURO in 2010), see *ibid.*

[85] Art. 1 Décret relative á la création d'un Fond Social de l'Eau en Région Wallonne, 20 February 2003, in: *Moniteur belge*, 19 March 2003, also available at http://www.environnement.wallonie.be/Legis/eau/eafon002.htm; Armeni (2008), p. 3.

[86] Armeni (2008), p. 3; also see documentation at http://www.worldwaterforum4.org.mx.

[87] Art. D.233bis–D.233bis10 Code de l'eau (included by a 8 August 2008 décret), http://environnement.wallonie.be/legis/Codeenvironnement/codeeaucoordonne.htm.

[88] See Art. 2.2 Ordonnance réglementant la fourniture d'eau alimentaire distribuée par réseau en région bruxelloise, 8 September 1994, in: *Moniteur belge*, 29 September 1994.

blocks of water consumption per person, followed by a fourth open block (above 60 m^3 per person per year) at a much higher (more than tripled) price.[89] Depending on household consumption, different tariffs are offered: a "vital" tariff (the most minimal one), a "social" tariff, a "normal" tariff, and a "comfort" tariff.[90] This way saving water is promoted, and basic water access is guaranteed. The region of Brussels has also created a social fund for water,[91] financed by a special tax per m^3, in order to enable poor households to pay their water bills. Water disconnection can only be legal after a preliminary investigation by welfare authorities and on the basis of an additional judicial decision.[92]

The Protection of the Right to Water through Belgian Courts

In 1996, the Belgian Court of Arbitration ruled famously on individual access to drinking water in its decision *A.s.b.l. Syndicat national des propriétaires et autres*.[93] The case concerned a regional ordinance of the Brussels region ruling that water distribution companies were forbidden to disconnect tenants who do not pay their water bills, but that the companies were entitled to account the water bill to the landlords instead.[94] An association of owners filed a petition to the Court of Arbitration in order to annul this respective norm. The owners argued that tenants were obliged to bear the costs of their consumption, claiming violations of the principles of equality (Art. 10) and non-discrimination (Art. 11) of the Belgian constitution. The court, however, refused to annul the norm in question. The arbitrators established that, although it was the tenants' primary obligation to pay their water bills, access to a minimum of drinking water and sanitation was a fundamental right, which must be guaranteed, even if tenants were unable to pay their bills. Through a form of obligation based on solidarity, owners were obliged to pay the water bills for their tenants in exceptional cases.[95] The court interpreted the

---

[89] See Art. 38 Ordonnance établissant un cadre pour la politique de l'eau, 20 October 2006, http://www.bruxellesenvironnement.be/Templates/download/20061020_Ord_cadre_Eau.pdf?langtype=2060; see also HydroBru's website (the local water provider of Bruxelles) at http://www.ibde.be/index.cfm?Content_ID.861833548; also see Armeni (2008), p. 3.

[90] See calculations at the webpage of *HydroBru* at http://www.ibde.be/index.cfm?Content_ID=861833548; see also Boury (2008), p. 5.

[91] See at the webpage of *HydroBru*, at http://www.ibde.be/index.cfm?Content_ID=998162340. See also Smets (2006b), p. 86; *id.* (2008a), p. 94.

[92] Armeni (2008), p. 3.

[93] Belgian Court of Arbitration, *A.s.b.l. Syndicat national des propriétaires et autre*, Case No. 9/1996, 8 February 1996, in: *Moniteur belge*, 1996(02)00035, quoted after Armeni (2008), p. 4.

[94] Art. 3.2 and 5.2 Ordonnance réglementant la fourniture d'eau alimentaire distribuée par réseau en région bruxelloise, 8 September 1994, in: *Moniteur belge*, 29 September 1994.

[95] Belgian Court of Arbitration, *A.s.b.l. Syndicat national des propriétaires et autre*, Case No. 9/1996, 8 February 1996, in: *Moniteur belge*, 1996(02)00035, section I, par. III, 2, quoted after Armeni (2008), p. 4.

ordinance in a way that every resident of a house, once connected to the distribution network, has the right to receive water for basic consumption. This guarantees best conditions for implementing the public task of water distribution and respects at the same the absolute right to a drinking water supply.[96]

Other Belgian courts followed this strong approach to the right to water. The Arbitration Court of the Commune of Wemmel[97] in 1998 explicitly promoted the idea that a subjective right to water for everyone was inherent in the right to a healthy environment contained in Art. 23 of the Belgian constitution. The case concerned the before mentioned legal framework for drinking water in the Flemish region. This framework allowed for a minimal amount of 15 m^3 of drinking water free of charge for everyone and progressive pricing. In 1998, a municipal drinking water company opposed this regulation, arguing that it interfered with the competence of the municipal water company over the determination of water prices. In its judgment, the Court of Arbitration stated that Flanders possessed the right to introduce this free basic amount, as the region had exclusive competence concerning water production and distribution including pricing.[98] The Court found that the region's regulation pursued the legitimate aim to implement the right of everyone to have access to a minimum of drinking water; a right protected under Art. 23 of the Belgian constitution. The progressive pricing in the region promoted rational and moderate water consumption, which was necessary to realize the right to enjoy a healthy environment: a free basic supply of water would cause a relative increase in water prices for extensive consumption and thus create incentives to save water for big users.[99]

Analysis of the Protection of the Right in Belgium

Belgian laws offer a very distinct approach to the right to water. Though being designed differently on the levels of the three regions Flanders, Wallonia and Brussels, a subjective right to water is effectively granted to all Belgians.

However, the right is not yet recognized as an individual entitlement at the national level, as the process to recognize the right to water as a fundamental constitutional right has been held up. In 2006, the Belgian Senate asked for an opinion of the *Federal Council of Sustainable Development* about the possible inclusion of the explicit recognition of the right to water in the constitution. The legal opinion of the council was split[100]: while some members argued that the

---

[96] *Ibid.*, par. B.5.

[97] Belgian Court of Arbitration, *Commune de Wemmel*, Case No. 36/1998, 1 April 1998, in: *Moniteur belge*, 24 April 1998, quoted after Armeni (2008), p. 5.

[98] *Ibid.*, par. B.4.2, 13.

[99] Armeni (2008), p. 4.

[100] Conseil Fédéral du Développement Durable, *Avis sur la révision de la constitution concernant le droit à l'eau*, CFDD Info Bulletin No. 26, March 2006, http://www.cfdd.be/DOC/pub/info/i27f.pdf.

growing importance of access to water made it necessary to recognize the right explicitly in the constitution, other members argued that in Belgium access to water of high quantity and quality was ensured within a solid ordinary legal framework. The implicit recognition of a right to water through interpretation of the right to health and the right to live in a healthy environment of Art. 23 was considered sufficient. As a result, the decision on the constitutional amendment has still not been made by the Belgian Parliament. This is a regrettable delay, if not an even more regrettable change of opinion. An explicit and independent right in the constitution would increase the coherence of the national system, and give the right an outstanding official recognition, ensure better implementation, and strengthen its visibility in the national legal order. However, given the recurring crisis of the Belgian government, torn between the different interests of the three regions,[101] one might doubt whether a consistent political line will be found in the future.

For the time being, the regions remain the main protectors of a right to water. Although the laws of the regions slightly differ, they all share a visible social concern for the realization of the right. Most of the legal provisions at the regional level emphasize the social dimension of the right to water: a free water quota, the ban on water disconnections and the progressive water prices all point in this direction. It is praiseworthy that the regions have taken the initiative to include as concrete as possible provisions into their legislation, and that the water providers are held to account in making their water pricing transparent. Water access is guaranteed to every person in Belgium, under conditions that are financially reasonable for each citizen. Careful and economical use of water is rewarded by a progressive water pricing system. The minimum amount of water necessary for survival is free (in Flanders) or at least very affordable (in the other regions). Fanciful waste of water is punished by high prices. This makes the protection of the scarce and precious source of water effective and emphasizes the social component of water as a common good. This strong legislative approach is accompanied and fortified by courts that are determined to give priority of the right to water over other rights, as the case of the Arbitration Court of the Commune of Wemmel shows. The courts very explicitly derive a right to water from Art. 23 of the Belgian constitution, and interpret regional water laws in the light of this provision.

The protection and the promotion of the right to water in Belgium must altogether be considered as very robust: a constitutional recognition (or, even if eventually not realized, at least a strong discourse about such recognition), the strong social concretization of the right to water through the regional laws, and a strong judiciary giving effect to the right where it is challenged putting its importance above other values and rights, like that of landlords and water companies.

---

[101] See for instance, Castle (2011); Müller (2011); or "Belgian politics in crisis as Prime Minister Yves Leterme quits again", in: *The Sunday Times Online*, 23 April 2010, http://www.thetimes.co.uk/tto/news/world/europe/article2487065.ece.

2.1.1.3 The Right to Water in France

French water laws have recently been renewed in light of a right to water, offering outspoken protection for the right under the French legal framework.

The Protection of the Right to Water in French Law

Most noticeably, through the introduction of the *Law on Water and Aquatic Environments*[102] ("Loi sur l'eau et les milieus aquatiques", LEMA) of 30 December 2006, France has provided a new dimension to its protection of the right to water. The law is designed to modernize the legislative framework for water management and to improve water quality. It is the French reaction to the EU Water Framework Directive, in order to meet the target of a good ecological status of surface water, set for 2015 by the Directive.[103] Art. 1 of the LEMA clarifies that the use of water is a right of everyone, and that every person has the right to have access to drinking water for drinking and hygiene under economically acceptable and universal conditions.[104]

This recent law is only the peak of several legislative acts used to promote the right to water in France over the last 15 years. Already in the beginning of the 1990s, France enacted a *Law on Water*,[105] later embedded in the *Environmental Code*.[106] Therein, France recognized water as a collective heritage of the French nation, and stated that the further development and protection of its use constituted a general interest.[107] In 2005, the law "*Oudin-Santini about international assistance*" added an international aspect to the thus-far national focus on water.[108]

---

[102] Loi n° 2006-1772 sur l'eau et les milieux aquatiques (LEMA), 30 December 2006, http://www.legifrance.gouv.fr/affichTexte.do?cidTexte=JORFTEXT000000649171.

[103] European Parliament and Council of the European Union, *Directive 2000/60/EC of the European Parliament and of the Council of 23 October 2000 establishing a framework for Community action in the field of water policy*, OJ 2000 L 327, 1.

[104] Art. 101 Loi n° 2006-1772 sur l'eau et les milieux aquatiques (LEMA), 30 December 2006, http://www.legifrance.gouv.fr/affichTexte.do?cidTexte=JORFTEXT000000649171: "Within laws, regulations and established rights, the use of water is for all and every physical person has the right of access to drinking water for nutrition and hygiene at affordable price."

[105] Loi n° 92-3 sur l'eau, 3 January 1992, http://www.legifrance.gouv.fr/affichTexte.do?cidTexte=JORFTEXT000000173995.

[106] Art. L210-1 ff. Code de l'environnement, http://www.legifrance.gouv.fr/affichCode.do?cidTexte=LEGITEXT000006074220.

[107] Loi n° 92-3 sur l'eau, 3 January 1992, http://www.legifrance.gouv.fr/affichTexte.do?cidTexte=JORFTEXT000000173995; Art. L210-1 Code de l'environnement, http://www.legifrance.gouv.fr/affichCode.do?cidTexte=LEGITEXT000006074220.

[108] Loi n° 2005-95 relative à la coopération internationale des collectivités territoriales et des agences de l'eau dans les domaines de l'alimentation en eau et de l'assainissement, 9 February 2005, http://www.legifrance.gouv.fr/affichTexte.do?cidTexte=JORFTEXT000000628190&dateTexte=&categorieLien=id.

Art. 1 and Art. 2 of that law authorized *inter alia* the French water agencies to enact projects in the field of international assistance and cooperation with respect to water.

With regard to water *availability*, the LEMA provides that minimum access to drinking water is guaranteed under all circumstances.[109] This includes priority of use as drinking water over other uses, and a right to take water from natural sources or public fountains for personal use without further authorization. More specific rights include the right to receive a minimum amount of water even in the case of crisis or interruption of service,[110] typically provided in France by dispensers such as "borne-fontaines" or "col-de-cygnes". Employees have, under certain circumstances, the right to receive drinking water for free at their workplace.[111]

Another aspect of the right to water which French law explicitly addresses is the *quality* of drinking water. Generally, the quality of drinking water is to be in conformity with the norms covering the protection of public health. Potable water cannot contain a concentration of micro-organisms, parasites or other substances that constitute a danger for human health.[112] Water quality must be surveyed by the public or private institutions responsible for the distribution of water. Under the revised version of the Code on public health, from 2013 onwards, anyone selling property destined for housing will have to provide a minimum standard of sanitary facilities.[113]

With respect to economic *affordability*, the French State recognizes that increasing water prices can be a serious obstacle to the realization of the right to water, and only charges a reduced value-added tax ("taxe à la valeur ajoutée") of 5.5 % on water.[114] Due to the current economic situation of France, the reduced tax rate was partly raised to 7.0 % on 1 January 2012.[115] The sale of water, however, was not affected by the amendment but continues to benefit from the 5.5 % reduced rate.[116]

---

[109] Art. L210-1 Code de l'environnement, http://www.legifrance.gouv.fr/affichCode.do?cidTexte=LEGITEXT000006074220.

[110] Art. 6 par. 1 Loi n° 2004–811 de modernisation de la securityé civile, 13 August 2004, http://www.legifrance.gouv.fr/affichTexte.do?cidTexte=JORFTEXT000000804612&dateTexte=&categorieLien=id.

[111] Art. R4225-2 ff. Code du travail, http://www.legifrance.gouv.fr/affichCode.do?cidTexte=LEGITEXT000006072050&dateTexte=20130301.

[112] Art. L1321-1, Art. R1321-23 Code de la santé publique, http://www.legifrance.gouv.fr/affichCode.do?cidTexte=LEGITEXT000006072665&dateTexte=20130301.

[113] *Ibid.*, Art. L1331-11-1.

[114] Art. 1 and 57 Loi n° 2006-1772 sur l'eau et les milieux aquatiques (LEMA), 30 December 2006, http://www.legifrance.gouv.fr/affichTexte.do?cidTexte=JORFTEXT000000649171; Art. L2224-12-3 Code général des collectivités territoriales (CGCT), http://www.legifrance.gouv.fr/affichCode.do;jsessionid=B34F48CD3214420710FB123903667168.tpdjo17v_2?cidTexte=LEGITEXT000006070633&dateTexte=20130301.

[115] Art. 13 Loi n° 2011–1978 de finances rectificative pour 2011, 28 December 2011, http://www.legifrance.gouv.fr/affichTexte.do;jsessionid=B34F48CD3214420710FB123903667168.tpdjo17v_2?cidTexte=JORFTEXT000025045613&dateTexte=20130301.

[116] *Ibid.*

Water is (still) accessible for an economically acceptable price. In November 2012, France announced plans to increase taxation in order to improve economic competitiveness, but to decrease the reduced tax rate to 5.0 %.[117]

The disconnection of water services in the case of continued non-payment has for a long time not explicitly been forbidden in French law. However, some regional water laws of the bigger cities[118] go in this direction. Additionally, after having proved their inability to pay for water services, households can, under certain circumstances, receive special financial aid to enable them to pay their water bills.[119] Under this right, 12 million EURO were distributed amongst 60,000 poor households in 2008,[120] financed by private water suppliers, communes and public general councils ("counseils généraux").[121] Finally, certain *information*-rights for consumers have recently been established under French law.[122] The public has the right to consult the annual reports that water suppliers have to give to the mayors of supplied local communities. These reports have to contain information on the development of water quality and on the prices of drinking water and sanitation.[123]

The Protection of the Right to Water through French Courts

French judges have also held a protective hand over some aspects of the right to water. Especially, the issue of disconnection of water services has been a topic that

[117] See "France to raise VAT to fund company tax breaks", *BBC News*, 6 November 2012, http://www.bbc.co.uk/news/business-20218792; Parienté (2012).

[118] See for instance Art. 21-3 Parisian Réglement de la distribution publique des eaux, 11–12 February 2013, http://www.eaudeparis.fr/page/abonne/demarches/reglement-duservice-public-de-l-eau?page_id=242; Art. 43 Nantes *Règlement du service de distribution d'eau potable*, 30 April 2004, http://www.cmaintenant.eu/files/file/reglement_du_service_de_distribution_d_eau_potable.pdf.

[119] Art. L115-3 Code de l'action sociale et des familles, http://www.legifrance.gouv.fr/affichCode.do?cidTexte=LEGITEXT000006074069; Art. 65 Loi n° 2004-809 relative aux libertés et responsbilités locales, 13 August 2004, http://www.legifrance.gouv.fr/affichTexte.do?cidTexte=JORFTEXT000000804607&dateTexte=&categorieLien=id. See also Loi n° 2011-156 relative à la solidarité dans les domains de l'ailimentation en eau et de l'assainissement (improving financing), 7 February 2011, http://www.legifrance.gouv.fr/affichTexte.do;jsessionid=B34F48CD3214420710FB123903667168.tpdjo17v_2?cidTexte=JORFTEXT000023557205&dateTexte020130301.

[120] Smets (2012), p. 9.

[121] *Ibid.*; see also "Statement by the Permanent Mission of France to the Office of the United Nations High Commissioner on Human Rights", on the occasion of the *Study on human rights obligations related to equitable access to safe drinking water and sanitation*, http://www2.ohchr.org/english/issues/water/contributions/France.pdf, p. 7.

[122] For instance Art. L1321-9 Code de la santé publique, http://www.legifrance.gouv.fr/affichCode.do?cidTexte=LEGITEXT000006072665&dateTexte=20130301.

[123] Art. L2224-1 ff. Code général des collectivités territoriales (CGCT), http://www.legifrance.gouv.fr/affichCode.do;jsessionid=B34F48CD3214420710FB123903667168.tpdjo17v_2?cidTexte=LEGITEXT000006070633&dateTexte=20130301.

different French courts have had to deal with on several occasions. The "Tribunal de Grande Instance" (TGI) of Avignon already ordered—in the mid 1990s—the reconnection of a household with four children to water services.[124] It argued that the detention of water, being an essential element for the life of a family, was a very serious matter and a risk for the family's health. The TGI of Privas added in a similar case, that the partial non-payment of water bills could not justify the disconnection of water services, and repeated that water was an essential element to life.[125] According to the TGI of Roanne, judges would have to decide from case to case whether a disconnection of water services constituted an evidently illicit breach of the consumers' right to water.[126] The TGI of Meaux ordered a water supplier to reconnect a caravan, which was used as a residence, to water services,[127] until it was decided in a (later) final judgment whether the caravan would have to be removed from where it was. The court stated that the request to reconnect a household to water services was not only an expression of property rights, but also of the inalienable right of human dignity and was a part of the public services essential for human life.

Apart from these cases of disconnection from water services, cases of water quality have also arisen before the French courts. In 1995, the "Tribunal d'Instance" of Guincamp ordered a water supplier to pay damages of more than 250,000 Francs to 176 households, for having provided them with water that contained more than 50 mg nitrate per liter for a period of longer than half a year. The supplier, for his part, received compensation through a judgment of the administrative court of Rennes, which found that the French State had failed to provide adequate surveillance of water quality.[128] In a similar case, the Court of Appeal of Rennes gave consumers the right to seek financial compensation from the water supplier for having to buy water bottles, as the quality of the tap water had been unacceptably low (again at a rate of more than 50 mg nitrates per litre).[129]

---

[124] Grand Instance Court of Avignon, *François X and the Union Fédérale des Consommateurs d'Avignon v. Société Avignonnaise des Eaux*, Case No. 1492/95, 12 May 1995, http://www.cace.fr/jurisprudence/rets/eaupression/tgi12051995.html.

[125] Grand Instance Court of Privas, *Compagnie de services et d'environnement v. Association des consommateurs de la Fontauliére*, Case No. 9800223, 5 March 1998, http://www.cace.fr/jurisprudence/rets/acf/tgi05031998.html.

[126] Tribunal de Grande Instance (District Court) of Roanne, *Revue CLCV*, No. 97, 11 March 1996.

[127] Tribunal de Grande Instance (District Court) of Meaux, *xxx. v. xxx.*, 28 February 2001, in: *Droit Monde*, No. 37–38, 2004, p. 77, quoted after COHRE (2008), p. 295.

[128] For both cases see Miquel (2013).

[129] Cour d'Appel of Rennes, Decision from 2003, quoted after OHCHR, "Statement by the Permanent Mission of France to the Office of the United Nations High Commissioner on Human Rights", 2007, http://www2.ohchr.org/english/issues/water/contributions/France.pdf, p. 9.

Analysis of the Protection of the Right in France

The legal recognition of the right to water in France has reached a similar level to the protection of the right to water in Belgium. In particular the water law of 2006 highlights the importance that France attaches to the issue of individual access to water. However, in contrast to the Belgian example, where the provinces provide the strongest protection of the right, in France this is the case mainly at the federal level, paired with the local municipalities. Given the centralized organization of the French Republic, this comes as no surprise. The right can be most efficiently addressed on that level which overall offers the strongest legal protection.

Depending on the circumstances, a technocratic legal approach can sometimes be more efficient than one that follows a heated and politicized debate. The example of the French right to water must be seen in this light. Without heated discussions for a constitutional amendment like in Belgium, the right has entered the legal stage successfully and effectively through the "backdoor" of ordinary laws. Given that some of the world's leading water companies are headquartered in France,[130] the influence of the water-lobby in France, paired with the view of water as an at least partly economic good, might have been too strong for the discussion on a constitutional amendment. However, recognition of constitutional status carries high symbolic value and would make it easier for judges to rule in favor of the right, where it clashes with other constitutional or international values or rights.

Regarding its normative content, the French recognition covers the most essential elements of the right to water, in particular accessibility, affordability and acceptable quality, but it still has some drawbacks. Firstly, only in some cities are disconnections explicitly illegal under all circumstances. In most cities and municipalities, however, disconnections can be justified under certain circumstances. Secondly, the municipalities have to provide water generally to all persons, but not always to people who live far away from urban centers.[131] This approach fails to recognize that it is often exactly these groups, in rural areas, that would need the benefits of a right to water the most. Thirdly, the concept of affordability is not yet sufficiently defined.[132] There is social assistance available for the lowest income households; yet middle-income households also must have their right ensured that water prices do not eat up a major part of their income.

The judiciary has ruled in many cases in favor of the right, where it was challenged in the ordinary courts. In particular, in cases in which basic water supply was cut off completely, the courts have been prepared to order its immediate reinstallation. The courts have on several occasions linked the right to water with

---

[130] Hall and Lobina (2012). See also the websites of some of the biggest French water corporations, Groupe SAUR, http://www.saur.com; Suez Environnement, http://www.suez-environnement.com; and Veolia Environnement, http://www.veolia.com.

[131] Art. L2224-7-1 Code général des collectivités territoriales (CGCT), http://www.legifrance.gouv.fr/affichCode.do;jsessionid=B34F48CD3214420710FB123903667168.tpdjo17v_2?cidTexte=LEGITEXT000006070633&dateTexte=20130301; also see Smets (2008b), p. 718.

[132] Smets (2008b), p. 718.

essential other rights (the right to life or the notion of human dignity) and ranked the right to water above competing other rights (e.g. rights of landowners or water suppliers).

Altogether, France is an example of a State that has credibly brought its legal situation at the national level in line with both its external political efforts and its international obligations. On the international level France has been a supporter for the promotion of the right to water over recent years[133] and was largely in favor of the EU Drinking Water Directive.[134] The LEMA has given the right recognition in the national legal order, and taken up the most pressing questions with regard to the right to water—quality, affordability and accessibility. That this protection is still incomplete is regrettable; yet, France's efforts are visible. As a water-abundant country, for France, to make commitments of this kind is easier and less costly than for many other States. How serious France's commitment to a right to water is, must be judged in the future also by the way it influences the big French water corporations in its international efforts. This is a variation in terms of an *international* obligation to protect[135]: in the case of France, it might be one of the most crucial ones in order to credibly show commitment to the human right to water.

2.1.1.4 Comparison of the Three European Approaches

As we have seen, the German, the Belgian and the French approach to the right to water are quite different. Whereas in Germany the legal base for the right to water can be found in general principles as the principle of the social State or the principle of "Daseinsvorsorge", Belgium and France have created differentiated legal orders for the protection and the promotion of the right. Belgium has done so through legislation on the regional level; France, coherent with its centralistic structure, has relied upon a federal law.

Although undoubtedly general consensus exists about the goal of universal access to water in Europe, the right to water is seen from very different perspectives in the three legal orders. Whereas in the German system the emphasis is put onto the responsibility of the State (arguably as the counterpart of its given privileges), Belgium and France take a rights-based approach, where the citizens' concrete entitlements take center stage. Germany adopts an approach based on principles that have more an objective focus on the State, whereas Belgium and France emphasize the right to water as a subjective right. The crucial question then arises

[133] See Smets (2006a), pp. 1–2. See also the statement of the French representative in the context of the 2010 UNGA Resolution on the human right to water and sanitation, UNGA, *Summary of the 108th plenary meeting of the 64th General Assembly*, A/64/PV.108, 28 July 2010, p. 14.

[134] Council of the European Union, *Council Directive 98/83/EC of 3 November 1998 on the quality of water intended for human consumption* (European Drinking Water Directive), OJ 1998 L 330, 32.

[135] See Sect. 4.3.3 of this book on the right to water and its international obligations.

as to which of these models is preferable and whether the different approaches make any difference in practice?

The first difference is who in fact shapes the right and fills it with concrete content. In Belgium and in France this task is largely performed by the legislator that enumerates concrete obligations and entitlements in the law. In Germany, the situation is the contrary. It is the judges who have the duty to fill the right to water with a concrete content. The broad principles, in conjunction with fundamental rights, that constitute the basis for the right to water in the German domestic legal system—namely the principle of the social State and the "Daseinsfürsorge", and the notion of human dignity—are fairly vague and do not provide a clear orientation for the courts. Thus, in Germany the protection of the right to water is more judge-made, while the protection is largely made by the legislator in Belgium and France (while also in these countries supported by an active judiciary to fill gaps where necessary).

One might not categorically prefer the one approach over the other. However, both approaches speak to two different paths for legitimization. The legislator is directly democratically legitimized through elections and votes. However, judges also have strong legitimacy for their action. Judges of highest national courts usually hold personal legitimacy, as their election or appointment regularly involves the participation of directly and democratically legitimized organs. The chain of democratic legitimacy in the case of judges is therefore prolonged, without being completely disconnected. On the other hand the *factual* legitimization of judges stems most often from their given tasks in national constitutions, emphasizing the principle of division of powers as an expression of "checks and balances", regularly provided by the constitution itself.[136]

An important question then of course is whether the courts can go beyond the protection that is explicitly offered by the positive law. The actual wording of the text often is not opposed to such an approach. The positive law can often be seen as a guaranteed minimum of legal protection, which does not exclude further protection offered or interpreted by judges. However, if the legal wording (as in the case of the Belgian and French water laws) already provides concretely-worded protection that is also far-reaching in scope (e.g. amounts of free water; size of different blocks of water prices, etc.), courts might be reluctant to go beyond this level of protection. They might rather accept the decision of the legislator. In this regard the German model of not stating particular entitlements and duties is more flexible, while the French, and particularly the Belgian, approach is more concise and thus in some respect more reliable for citizens.

Thirdly, the Belgian and the French subjective-rights approach stays abreast of the relation of national rights to water to the idea of a human right to water. Human rights are rights that are possessed by all persons just by virtue of being human. This humanistic focus should also be reflected in national laws. It is from the value of human dignity that these rights ultimately stem in their very core, not from State

[136] See generally on this topic, Goeters (2008), pp. 62 ff.; Möllers (2005), pp. 398 ff.

responsibility or the nature of the State order. Even in the absence of any State order (e.g. in cases of failed States) human rights must still be valid; for a right typically challenged in times of State crisis and emergency, such as the right to water, this should even more be the case. Therefore, the approach in reflecting the urgent human need for water on the national level in the form of a subjective right, instead of as part of general State-focused principles, appears more coherent and the overall preferable concept in combining international and domestic law.

Whether this recognition reaches up to the constitutional level (as for a long time planned in Belgium) or whether it is limited to ordinary laws (as in France and ultimately in Belgium) is a question the national legislator has to decide. We will see later that such decision might be crucial in determining whether the right takes precedence over other (international) rights.[137] Apart from this question, it would be problematic, if the right to water was exclusively mentioned in the constitution, without concrete provisions in the laws. Had the Belgian constitutional amendment been successful (or if it is in the future), it would have combined concrete legal entitlements in the ordinary law with a visible political commitment in the constitution. This could have been considered an optimal approach to a right to water. However, ultimately the best practice for a country must be decided from country to country—taking into account national legal as well as cultural traditions.

However, apart from such culturally relativist conclusion, one should remember to mention what all three countries have in common. The judiciary played a crucial role for the protection of the right to water to take concrete form. Wherever there was a gap, the judges were prepared to fill it: be it by creating such a right in the first place through linking constitutional principles and fundamental rights, as in the German case, or by upholding the legally explicitly recognized right against competing legal values, as in the case of France and Belgium. It is not an overstatement to say that the right to water in the examined European legal orders is largely judge-made. This is an important lesson to keep in mind when turning our attention to the origins of a (potential) human right to level at the international level.[138]

2.1.2 The Right to Water in European Law

In the interlinked legal systems in Europe the legal protection of rights in the Member States must always be seen with regard also to the legal protections offered on the European level, namely in the system of the EU and under the regime of the ECHR and the ESC.

[137] See Sect. 3.4.4 of this book.

[138] See Sect. 2.3 of this book.

2.1.2.1 The Right to Water in the Law of the European Union

The topic of water itself has a significant EU dimension. Several water directives cover a large number of water topics such as issues concerning groundwater,[139] surface water,[140] or drinking water.[141] In the latter it is stated that the Member States shall ensure that any supply of water intended for human consumption which constitutes a potential danger to human health is prohibited or its use restricted; any action necessary must be taken to protect human health.[142] This shows the awareness within the union of the link between water and overall common values like public health.

Most importantly, in 2000 the EU established a comprehensive framework for Community action in the field of water policy.[143] In its preamble, par. (1), the Member States declared that water is not a commercial product like any other, but rather a heritage which must be protected, defended, and treated as such. Art. 1 continues to state that the Member States are to contribute to the provision of the sufficient supply of good quality surface water and groundwater as needed for sustainable, balanced and equitable water use.

Most of these directives belong to the field of European environmental law or consumer protection, as the union has far-reaching competences in these fields. They aim at the protection and improvement of water quality at an abstract level. Surely, however, behind these values lies the aim of protecting the individual. However, this is a purely non-rights orientated approach. The Charter of Fundamental Rights of the European Union (CFREU),[144] finally legally binding since December 2009 via its inclusion into the Lisbon Treaty,[145] also does not include any mention of a subjective right to water.

With regard to political statements, the European Parliament took action in 2003 and emphasized the human rights character of the right to water and advocated

[139] Council of the European Communities, *Council Directive 80/68/EEC of 17 December 1979 on the protection of groundwater against pollution caused by certain dangerous substances*, OJ 1980 L 20, 43.

[140] Council of the European Communities, *Council Directive 75/440/EEC of 16 June 1975 concerning the quality required of surface water intended for the abstraction of drinking water in the Member States*, OJ 1975 L 194, 26.

[141] Council of the European Union, *Council Directive 98/83/EC of 3 November 1998 on the quality of water intended for human consumption* (European Drinking Water Directive), OJ 1998 L 330, 32.

[142] *Ibid.*, preamble (26) and Art. 1 par. 2.

[143] European Parliament and Council of the European Union, *Directive 2000/60/EC of the European Parliament and of the Council of 23 October 2000 establishing a framework for Community action in the field of water policy*, OJ 2000 L 327, 1; Faßbender (2001), p. 241.

[144] CFREU, 18 December 2000, OJ C 364/1 (2000), 1.

[145] Art. 6 TEU, OJ 2012 C 326, 13.

European action in this field.[146] Even more, in 2012, one of the first *European Citizens' Initiatives* ("Water is a Human Right") invited the European Commission to initiate legislation to implement a human right to water and sanitation,[147] and has in the meantime reached the necessary number of signatures[148] (one million signatures from at least a quarter of the EU Member States within 12 months[149]). The new instrument of the European Citizens' Initiative is regulated in Art. 11(4) TEU and Art. 24(1) TFEU; the concrete procedure was refined by regulation 211/2011.[150]

Both of these events marked first political steps in the direction of the promotion of a subjective right. However, the 2003 declaration was primarily directed at developing countries outside the EU. As far as the EU itself was concerned, the resolution was meant to call for a European strategy in terms of its water development policy. The resolution was neither directed at the Commission nor at the Member States to compel them to establish an explicitly recognized right to water in the legal framework of the Union or the Member States.

Such a direct appeal to the Commission is now made by the new European Citizens' Initiative. The prospects of this brand new instrument remain, for the time being, uncertain. Despite the fact that the necessary number of signatures was easily reached by the initiative on a right to water, it is important to note that the initiative can only *call* on the Commission to propose a new legal act. It cannot obligate the Commission to do so. This follows on the one hand from the clear wording of Art. 11(4) TEU (the initiative can "invite" the Commission to initiate legislation); it follows on the other hand from the well-established principle of the Commission's monopoly to initiate legislation, which was not meant to be challenged by the new Art. 11(4) TEU.[151] However, given the enormous effort that it takes the initiators and in light of the *telos* of Art. 11(4) TEU to intensify the democratic link between the Commission and the European people,[152] one will have to assume the duty of the Commission to at least consider and deliberate the Citizens' Initiative.[153] A direct right to water can, however, clearly not be derived from the, albeit politically very successful, European Citizens' Initiative on the human right to water.

---

146 European Parliament, *Resolution on the Commission communication on water management in developing countries and priorities for EU development cooperation*, 4 September 2003, OJ C 76 E, 430.

147 Background information to the European Citizens' Initiative are available at http://ec.europa.eu/citizens-initiative/public/welcome?lg=en.

148 See for the status of signatures the official website at http://www.right2water.eu. See also European Commission, *European Citizens' Initiative hits 1 million signatures*, Press release IP/13/107, 11 February 2013.

149 Art. 11(4) TEU, OJ 2012 C 326, 13; Art. 24(1) TFEU, OJ 2012 C 326, 47; Art. 5(5) European Parliament and Council of the European Union, *Regulation (EU) No. 211/2011 of the European Parliament and of the Council of 16 February 2011 on the citizens' initiative*, OJ 2011 L 65, 1.

150 *Ibid.*

151 Craig and de Búrca (2011), pp. 122–123; Nettesheim (2012), Art. 11, par. 26; Vedder (2012), Art. 11, par. 3.

152 Jaag (2012), p. 317.

153 Geiger (2010), Art. 11, par. 3.

There is another notable provision with regard to water in the Lisbon Treaty. According to Art. 192(2)(b) TFEU, the only norm that deals with the issue of water directly, European regulations and directives establishing measures that affect the quantitative management of water resources or their availability must be adopted by an *unanimous* vote by the Council. This is a clear derogation from the usual procedure applicable in EU environmental policy, which is to adopt legislation through a co-decision of the Council and the European Parliament.[154] This emphasizes the sensitive importance of the management of water resources for the Member States and at the same time the difficulties to internationalize issues of water supply. The exceptional requirement of unanimity shows that Member States remain unwilling to lose any rights over their internal water sources from their own grasp.

Not having found explicit European norms on a right to water, one has to turn to more general concepts, such as the services of general economic interest, which, as the Commission has explicitly stated, include the supply of water.[155] In the German versions of the communications from the Commission concerning the services of general economic interest, the term "Daseinsvorsorge" is used as a synonym for these services.[156] In addition to these general statements, the term of the services of general economic interest is mentioned in the Lisbon Treaty in Art. 14 TFEU and in Art. 106(2) TFEU. Although competence in the area of services of general economic interest generally lies in the hands of the Member States—which can be seen from the exception laid down in Art. 106(2) TFEU as well as from Art. 14(1) TFEU—Art. 14(2) TFEU interestingly suggests a new competence of the EU for establishing guidelines and conditions for ensuring the functioning of services of general economic interest. Apart from the texts of the treaties, the concept of services of general economic interest was mainly developed in the course of the deregulation of the public utility industry.[157] Through the introduction of competition in fields like telecommunication and post as well as electricity and gas, it was meant to be ensured that a basic supply of certain services for the population under acceptable price and quality conditions could still be upheld after the introduction of EU-wide competition.[158] Therefore, the EU developed, separately for the different economic sectors, the concept of the "universal service".[159] Regarding the experience of other fields within the area of services of general interest, it is apparent that the right to a basic supply of certain goods or services, including concrete entitlements, was always

[154] Art. 192(1) TFEU, OJ 2012 C 326, 47.

[155] European Commission, *Communication from the Commission: Services of General Interest in Europe*, COM(96) 443, 11 September 1996, OJ 1996 C 281, 3; European Commission, *Communication from the Commission: Services of General Interest in Europe*, COM(2000) 580, 20 September 2000, OJ 2001 C 17, 4

[156] The concept of "Daseinsvorsorge" in the German system, see Sect. 2.1.1.1 of this book.

[157] Ruge (2003), p. 353; Kämmerer (2004), p. 33.

[158] Ruge (2003), p. 354.

[159] See for instance European Parliament and Council of the European Union, *Directive 2002/22/EC of the European Parliament and of the Council of 7 March 2002 on universal service and users' rights relating to electronic communications networks and services (Universal Service Directive)*, OJ 2002 L 108, 51.

only established in parallel to liberalization in these areas. With the general competence to regulate water services still being in the hands of the Member States, it is not surprising that individual rights to drinking water are not (yet) laid down in EU law. In its current from, a subjective right for the individual cannot be derived from the principle of the services of general economic interest.[160]

Most relevant for the derivation of a right to water is also the CFREU,[161] which is a legally binding part of EU law since the Lisbon Treaty entered into force in 2009 (Art. 6 TEU and Art. 51 CFREU). Within the CFREU, several articles can be indirectly considered to be parent rights for a right to water. The right to human dignity (Art. 1 CFREU) and the right to life (Art. 2(1) CFREU) as well as the right to physical integrity (Art. 3(1) CFREU) cover the most essential claims with respect to water (necessary for bare survival); the rights to social security (Art. 34 CFREU, in particular Art. 34(3) CFREU) and the right to health (Art. 35 CFREU) add more specific claims with respect to water. These rights are complemented by the recognitions on access to services of general economic interest (Art. 36 CFREU) and environmental protection (Art. 37 CFREU).[162] The high number of these relevant norms within the regime of the CFREU highlights that it is not clear so far in European law which norms must ultimately be considered decisive as parent rights for a right to water. The right to water presents itself as a fragmented right, without clear origin, but with strong connections to many of the established rights.

2.1.2.2 The Right to Water in the ECHR and the ESC

European law in a broader sense includes the law of the ECHR and the ESC. An explicit right to water is not included in the ECHR.[163] The rights therein are mainly of a civil and political nature, which was one of the reasons for the development of the ESC. Only rarely did cases relating to water become relevant in the jurisdiction of the European Court of Human Rights (ECtHR) in Strasbourg; if so, this was mainly in the context of Art. 3 ECHR, the prohibition of torture and inhuman treatment.[164] The court has found on several occasions, most explicitly in the cases of *Elci and others v. Turkey*[165] and in *Ostrovar v. Moldova*,[166] that not providing

[160] Kämmerer (2002), p. 1043.

[161] CFREU, 18 December 2000, OJ C 364/1 (2000), 1.

[162] Most of these rights are the European equivalents to international human rights. In order not to duplicate discussions, see the linkage of the right to water to these rights in the discussion on deriving the right in international law at Sect. 2.3 of this book.

[163] Council of Europe, The European Convention for the Protection of Human Rights and Fundamental Freedoms, 4 November 1950, UNTS 213 (1950), 222.

[164] *Ibid.*, Art. 3: "No one shall be subjected to torture or to inhuman or degrading treatment or punishment."

[165] ECtHR, *Elci and others v. Turkey*, Application No. 23145/93 and 25091/94, Judgment of 13 November 2003, http://hudoc.echr.coe.int/.

[166] ECtHR, *Ostrovar v. Moldova*, Application No. 35207/03, Judgment of 13 September 2005, http://hudoc.echr.coe.int/.

prisoners with sufficient drinking water of acceptable quality or denying them regular access to bathing facilities constitutes inhuman treatment in the sense of Art. 3 ECHR. Occasionally, the court also linked access to water to the right to property. In the case *Zander v. Sweden*,[167] which concerned pollution of a drinking water well on the estate of the applicants through a nearby dump, the Court found that drinking water contamination can constitute a violation of the right to property. This finding is interesting as the court assumes that groundwater can be part of private property. However, *Zander v. Sweden* must be considered a rather special case that does not give any hint as to what the court's position with respect to a right to water in general is.

The ESC[168] and the Revised ESC,[169] which is not enforceable by a court, but subject to a supervisory mechanism under the European Committee of Social Rights (ECSR),[170] do not explicitly accept a right to water either. However, some provisions of the ESC could be linked to such a right.[171] Most prominently, Art. 11 ESC, the right to protection of health, states that the contracting parties should remove as far as possible the causes of ill health, and prevent as far as possible epidemics, endemics or other diseases. This will not be possible, if access to safe drinking water is not made available. Similar logic can be applied to Art. 31 of the Revised ESC, where State parties are obliged to promote access to housing of an adequate standard. According to the ECSR, housing under Art. 31 is considered adequate if, *inter alia*, a dwelling is safe from a sanitary and health point of view and if it possesses all basic amenities, including water.[172] Public authorities must also guard against the interruption of essential services such as electricity, telephone and water.[173]

2.1.3 Conclusion and Analysis

Within the law of the EU, a subjective right to water is hard to trace. Many norms could be treated as legal sources. European regulations dealing with the amelioration of water quality might also serve the interest of the individual to ensure supply of high-quality drinking water, but have, however, not (yet) led to a recognition of an individual right within the EU. The new *European Citizens' Initiative* on the

---

[167] ECtHR, *Zander v. Sweden*, Application No. 14282/88, Judgment of 25 November 1993, [1993] ECHR Ser. A No. 279B.

[168] Council of Europe, European Social Charter, 18 October 1961, CETS No. 035.

[169] Council of Europe, European Social Charter (revised), 3 May 1996, CETS No. 163.

[170] See *Rules of the European Committee of Social Rights*, 29 March 2004, http://www.coe.int/t/dghl/monitoring/socialcharter/escrrules/Rules_en.pdf; and ECSR (2012), pp. 18 ff.

[171] Some of these rights are the European equivalents to international human rights; see discussion of the linkage of the right to water to these rights at Sect. 3.4.2 of this book.

[172] ECSR (2006), Art. 31, p. 294, par. 695, 697, in reference to Conclusions 2003, France, p. 221.

[173] *Ibid.*, par. 699.

right to water and sanitation might set a (European) precedence for the establishment of a right following an initiative from civil society, but this still remains to be seen. So far, the EU has concentrated to engage itself in the promotion of a human right to water in the international scene. It would be a more coherent approach if the Union engaged itself at the same time in promoting a rights-based approach in the different Member States and establishing such a right in the Union itself, as requested by the Citizens' initiative.

Europe at present does not face severe problems regarding water supply. The Union should not, however, disregard the question of water access, for three main reasons. Firstly, as just mentioned, the Union could promote a right to water on the international level more effectively, if the Union and the Member States themselves had made a clearer and more visible commitment to a subjective right to water. Secondly, the Union should pre-emptively protect itself as times of water scarcity could also arise in Europe, especially in southern Europe, in particular with the prospects that the EU will expand into even dryer regions in the future, such as Croatia or even Turkey.[174] Thirdly, the right to water has various aspects, including economic affordability. This is even under the present conditions of water abundance a highly relevant question for the citizens of Europe. However, looking back to experiences in other areas of "services of general economic interest", a subjective right to water in the framework of the "universal service" will only be discussed (and potentially protected) where action is taken to liberalize service provision in this field.

In the framework of the ECHR and the ESC, the right to water is also not recognized explicitly. However, the court and the committee have linked some rights and established certain elements of a right to water. In the framework of the ECHR these are mainly cases involving the mistreatment of prisoners where the right to water comes into play. The court finds the deprivation of water an affront against humanity as the court regularly qualifies these cases as inhumane treatments under Art. 3 ECHR. Thus, the European judges emphasize the element of human dignity within the right to water. Within the catalogue of the ESC, especially the right to housing has close links to the right to water: according to the ECSR, adequate housing must always include a minimum supply of water of acceptable quality. However, given the very limited enforceability of the rights protected under the ESC and the Revised ESC, the benefits for the citizen of greater linkage between the right to housing and a right to water remain small.

Altogether, a subjective right to water has not much arisen before European judges and did not find much mention in relevant legal texts. With regard to the EU, this is mainly owing to two reasons: firstly, that the region of the EU is in big parts water-abundant, and secondly, because the European Union has few competences in the field of water, where national standards exceed standards for environmental protection, as particularly the unanimity requirement of

[174] See for an overview of countries seeking EU membership: "EU enlargement: The next seven", in: *BBC News*, 1 July 2013, http://www.bbc.co.uk/news/world-europe-11283616.

Art. 192(2)(b) TFEU emphasizes. This, in turn, is an expression of how much national States regard their internal water, and the provision thereof, as an internal affair. They are unwilling to share their water, even with their European friends and partners.

2.2 The Right to Water Outside of Europe

Different from in Europe, some non-European countries have given the right to water the status of a fundamental right in their constitutions. These countries include *inter alia*[175] Congo,[176] Ecuador,[177] Ethiopia,[178] Gambia,[179] Kenya,[180] Maldives,[181] South Africa,[182] Uganda,[183] Uruguay,[184] Venezuela,[185] Zambia,[186] and the two American States of Massachusetts[187] and Pennsylvania.[188] Notably, the

[175] See Smets (2006b), p. 44.

[176] Art. 48 Constitution of the Democratic Republic of the Congo, 2006, http://www.constitutionnet.org/files/DRC%20-%20Congo%20Constitution.pdf.

[177] Art. 3(1) and Art. 12 Constitution of the Republic of Ecuador, 2008, http://pdba.georgetown.edu/Constitutions/Ecuador/english08.html.

[178] Art. 90(1) Constitution of the Federal Republic of Ethiopia, 1994, http://www.africa.upenn.edu/Hornet/Ethiopian_Constitution.html.

[179] Art. 216(4) Constitution of the Republic of the Gambia, 1996, http://www1.umn.edu/humanrts/research/gambia-constitution.pdf.

[180] Art. 53(1)(d) Constitution of the Republic of Kenya, 2010, http://www.kenyalaw.org/klr/index.php?id=741.

[181] Art. 23(a) Constitution of the Republic of the Maldives, 2008, http://www.maldivesinfo.gov.mv/home/upload/downloads/Compilation.pdf.

[182] Constitution of the Republic of South Africa, 1996, http://www.info.gov.za/documents/constitution/1996/a108-96.pdf, Section 27.1(b).

[183] Art. XIV(b) (Preamble) Constitution of the Republic of Uganda, 1995, http://www.ugandaembassy.com/Constitution_of_Uganda.pdf.

[184] Art. 47 Constitution of the Republic of Uruguay, 1967 (last amended in 2004), http://www.parlamento.gub.uy/constituciones/const004.htm (Spanish version).

[185] Bolivarian Republic of Venezuela, 1999, http://venezuelanalysis.com/constitution.

[186] Art. 112 Constitution of Zambia, 1996, http://www.refworld.org/docid/3ae6b5610.html; and see also Art. 70(2) Constitution of Zambia (Draft), 2012, http://www.zambia.co.zm/downloads/draft_constitution.pdf.

[187] Art. XCVII Constitution of the Commonwealth of Massachusetts, 1780, http://www.malegislature.gov/Laws/Constitution.

[188] Art. 27 Constitution of the Commonwealth of Pennsylvania, 1978, http://sites.state.pa.us/PA_Constitution.html.

new 2012 constitution of Egypt[189] and the 2011 Transitional Constitution of South Sudan[190] guarantee the right to water. Also, the most recent (new) constitution of Zimbabwe, approved by an overwhelming majority in the referendum of March 2013,[191] now grants "the right to safe, clean and potable water".[192]

Progressive as some of these constitutions are (as South Africa's "young" constitution of 1996)[193] or as new as the commitment to a right to water in some constitutions is (as in Uruguay after its constitutional amendment in 2004),[194] one is tempted to assume a new general trend in national constitutional law towards a right to water. However, this would be a premature conclusion. The majority of States so far have not followed in the footsteps of countries such as South Africa or Uruguay. In the latter case the legal change was not even instigated by State authorities, but endorsed and compelled by a referendum based on different initiatives of social groups and non-governmental organizations (NGOs).[195]

2.2.1 *The Right to Water in South African Law*

South Africa has taken several steps to make the right to water concrete by implementing an effective access to water policy. With a proclaimed objective for every individual to have an adequate water supply,[196] South Africa has often been called one of the biggest advocates of a right to water, particularly when it

[189] Art. 68 Constitution of the Arab Republic of Egypt, 2012, http://niviensaleh.info/constitution-egypt-2012-translation. As an example for the critique on the constitution, see Amnesty International, "Egypt's new constitution limits fundamental freedoms and ignores the rights of women", 30 November 2012, http://www.amnesty.org/en/news/egypt-s-new-constitution-limits-fundamental-freedoms-and-ignores-rights-women-2012-11-30.

[190] Art. 35(2) Transitional Constitution of the Republic of South Sudan, 2011, http://www.sudantribune.com/IMG/pdf/The_Draft_Transitional_Constitution_of_the_ROSS2-2.pdf.

[191] See—also on reports of police harassment and intimidation—the articles "Zimbabwe approves new constitution", in: *BBC News*, 19 March 2013, http://www.bbc.co.uk/news/world-africa-21845444; and Shaw and Gotora (2013).

[192] Art. 77(a) Constitution of Zimbabwe (Final Draft), 2013, http://www.gta.gov.zw/index.php/documents/constitution-of-zimbabwe.

[193] Constitution of the Republic of South Africa, 1996, http://www.info.gov.za/documents/constitution/1996/a108-96.pdf; see Constitution's history of origin at http://www.sahistory.org.za/constitution-history; also see Francis (2005), p. 156.

[194] The right to water as a fundamental right was inserted in Art. 47 of the Uruguayan Constitution after a referendum on 31 October 2004; see Santos and Villareal (2006), p. 1.

[195] *Ibid.*, pp. 1–2.

[196] Human (2006), p. 83; Smets (2006b), p. 92.

comes to concrete national implementation.[197] Although Africa as a whole has an estimated 344 million people lacking access to improved drinking water sources and is one of the most water-stressed regions in the world,[198] South Africa has successfully started to reduce the number of people without access to safe drinking water by a number of around 15 million people in the last decade.[199] The South African protection and promotion of the right to water through legislation, and the role of its courts, therefore deserve a more detailed analysis.

2.2.1.1 Protection of the Right to Water through South African Legislation

South Africa's constitution of 1997 proclaims in section 27 par. 1 (b) that everyone has "the right to have access to [...] sufficient food and water", specifying in section 27 par. 2 that "the State must take reasonable legislative and other measures, within its available resources, to achieve the progressive realization of these rights." The right to water is only one of a whole list of economic, social and cultural rights that were included in the text of the new South African constitution, such as access to housing, food, health care, social security, education and a healthy environment.[200] The constitution certainly has a far-reaching commitment to socio-economic rights.[201]

One of the results of the constitutional instruction of concretization is the *Water Services Act* of 1997,[202] which states that everyone has a right of access to basic water supply and basic sanitation, and that every water services institution must take reasonable measures to realize these rights.[203] The overall objective of the Act is to assist local governments in their functions of water services provision and to look after the interest of consumers. Although the water service provider is generally given the right to discontinue services, if there is failure to comply with the reasonable conditions for such provision on the part of the customer,[204] procedures for the limitation or discontinuation of water services must be fair and equitable and may never result in a person being denied access to basic water services for non-payment.[205] Reasonable notice and the opportunity to explain oneself must be given.

---

[197] Mehta (2005), p. 237; Unknown (2007), pp. 1082 ff.

[198] WHO and UNICEF (2012), p. 3.

[199] Human (2006), p. 83.

[200] Constitution of the Republic of South Africa, 1996, http://www.info.gov.za/documents/constitution/1996/a108-96.pdf, see sections 24, 26, 27 and 29 respectively.

[201] Scott and Alston (2000), p. 214.

[202] Republic of South Africa, Water Services Act, Act. No. 108, 1997, http://www.ielrc.org/content/e9705.pdf.

[203] *Ibid.*, section 3(1) and (2).

[204] *Ibid.*, section 11(2)(g).

[205] *Ibid.*, section 4(3)(a) and (c).

In case of privatization, the local governments—the default service providers—have to ensure that their subcontractors comply with the right to water.[206] Based on the Water Services Act, several Ministerial Regulations exist which relate to compulsory national standards and measures to conserve water.[207] Amongst them, the relevant regulation defines the term basic water supply as a minimum of 25 L of drinking water per day per person, or 6 kL per household per month.[208]

Furthermore, the *National Water Act*[209] of 1998 states that all water, whether surface or underground, belongs to the nation.[210] The Minister of Water Affairs, acting on behalf of the national government as trustee of the water, has to ensure that South African water is protected, used, developed, conserved, managed and controlled for the benefit of all persons.[211] It also establishes the "reserve": the guarantee that all water resources are used to a certain degree for basic human needs, so that basic human needs enjoy priority over other water demands.[212] The National Water Act also establishes a system of progressive water tariffs. It authorizes different water tariffs based on the users' economic situations, as "social equity" is one of the decisive criteria for setting different tariffs.[213] According to the Free Basic Water Policy,[214] the government even provides 200 L of free drinking water per day to each poor household (6,000 L per month). According to the Department of Water Affairs (as of September 2010), 86.1 % of South African households and 86.55 % of households with low income received water for free.[215]

---

[206] Kok (2005), p. 280; Winkler (2010), p. 426.

[207] See South African Department of Water Affairs and Forestry, Regulations relating to Compulsory National Standards and Measures to conserve Water No. R509, 8 June 2001, http://cer.org.za/wp-content/uploads/2011/10/Regulations-relating-to-compulsory-national-standards-and-measures-to-conserve-water.pdf.

[208] *Ibid.*, Regulation 3.

[209] Republic of South Africa, National Water Act, Act No. 36, 1998, www.dwaf.gov.za/Documents/Legislature/nw_act/NWA.pdf.

[210] *Ibid.*, Preamble.

[211] *Ibid.*, chapter 1, 3(1).

[212] *Ibid.*, chapter 1, 1(1)(xviii)(a); also see Liebenberg (1998); Winkler (2010), p. 424; *id.* (2012), pp. 146–147.

[213] Republic of South Africa, National Water Act, Act No. 36, 1998, www.dwaf.gov.za/Documents/Legislature/nw_act/NWA.pdf, chapter 5, Financial Provisions, pre-text to 56(1).

[214] Free Basic Water Program of the South African Department of Water Affairs and Forestry, http://www.dwaf.gov.za/documents/fbw/qabrochureaug2002.pdf. On the economic effectiveness of this program, see Szabo (2010).

[215] Department of Water Affairs (2010), p. 26.

2.2.1.2 Protection of the Right to Water through South African Courts

The constitutional court, being the highest court in all South African constitutional matters, supported by the South African High Courts, has shown on several occasions an even more proactive approach than becomes already apparent in the laws.[216] In the water-jurisprudence of the South African courts, the *Manquele*-case, the *Grootboom*-case, the *Bon Vista*-case and the recent *Mazibuko*-case have all been landmark decisions, both positively and negatively, for the implementation of the right to water at the national level.

In the *Manquele v. Durban Transitional Metropolitan Council*-case[217] the applicant, an unemployed woman, who occupied premises with her seven children, sought a declaratory order that the water-discontinuation to her premises by a local council after non-payment was unlawful. She relied on her right to a basic water supply as laid down in the Water Services Act. After the respondent had argued that the term "basic water supply" of the Water Services Act, due to the lack of any further regulations, was in fact *without content*,[218] the applicant revised her argument and rather relied on section 27 of the South African constitution, which guarantees the "right to sufficient water to everyone". She argued that therefore a certain minimum of sufficient water supply must be read into the provision of the Water Services Act.[219] The court, however, rejected these constitutional arguments as they had been made too late.[220] It accepted the respondent's argument that the court had no guidance from the legislature or executive to enable it to interpret the content of the right to water of the Act.[221] Defining the quantity and quality of the applicants' entitlements, and thereby concretizing the term "access", were political tasks that, as judge Niles-Dunér in his judgment stated, "lie outside the purview of my role and function".[222] In sum, the court found the right to water in the South African regime "incomplete and accordingly unenforceable"[223] and rejected the order.

Secondly, in the *Grootboom*-case[224] applicant Irene Grootboom and her children, who lived in extreme poverty, had been evicted from their temporary homes. Mrs. Grootboom applied to the High Court and subsequently to the Constitutional

[216] Human (2006), p. 86.

[217] High Court (Durban and Coast Local Division) of South Africa, *Manqele v. Durban Transitional Metropolitan Council*, Judgment of 7 February 2001, 2002 6 SA 423 (D).

[218] *Ibid.*, 426 (G). At the time, the above mentioned Regulation had not been released.

[219] *Ibid.*, 427 (A).

[220] *Ibid.*, 427 (H&I).

[221] *Ibid.*, 427 (D&E).

[222] *Ibid.*, 427 (E).

[223] *Ibid.*, 427 (F).

[224] Constitutional Court of South Africa, *The Government of the Republic of South Africa and Others v. Irene Grootboom and Others*, Case No. CCT 11/00, Judgment of 4 October 2000, 2000 (11) BCLR 1169 (CC), 1185.

Court for an order requiring the government to provide them with adequate basic shelter and housing including the provision of services such as water, until they obtained permanent accommodation. The courts ruled against the local government for failing to fulfill its constitutional obligation regarding water. It confirmed that the constitution required the State to act positively[225] to give effect to constitutionally guaranteed social and economic rights; otherwise they would exist on paper only.[226] Not until socio-economic rights were afforded to all people, could citizens enjoy their other rights, namely human dignity, freedom and equality.[227] With respect to water and sanitation, the municipality was obliged to install latrines and ensure a regular supply with water for the inhabitants of the shantytown[228]; the government was to provide the municipality with a subsidy for this purpose. As section 39 of the South African constitution obliges the courts to consider international law as a tool of interpretation of the Bill of Rights,[229] the courts also referred to the right to adequate housing laid down in the International Covenant on Economic, Social and Cultural Rights (ICESCR), and the duties of States implicated therein.

Thirdly, in the *Bon Vista*-case,[230] the applicants sought interim relief for the reconnection of their water supply after not having paid the bills for some months. Unlike in the *Manqele*-case, the applicants in *Bon Vista* relied from the start on their constitutional right to water. The court in its decision held that the Local Council, as an organ of the State, may not take any measure that results in preventing existing access to water.[231] By disconnecting the applicants' water supply, the council had *prima facie* breached the applicants' constitutional right to have their access to water respected.[232] The court found that in the case of any disconnection, the procedure must be fair and equitable and has to provide for the opportunity to make representations.[233] Where a person proves they are unable to pay for basic water services,[234] services may not be discontinued. The Court also referred to jurisprudence of the ESCR-Committee,[235] as the court in *Grootboom* had done before it.

---

[225] *Ibid.*, par. 93.

[226] *Ibid.*, par. 20.

[227] *Ibid.*, par. 23.

[228] *Ibid.*, par. 4.

[229] Section 39(1) of the Constitution provides: "When interpreting the Bill of Rights, a court, tribunal or forum [...] (b) must consider international law."

[230] High Court (Witwatersrand Local Division) of South Africa, *Residents of Bon Vista Mansions v. Southern Metropolitan Local Council*, Case No. 01/12312, Judgment of 5 September 2001, 2002 (6) BCLR 625 (W).

[231] *Ibid.*, par. 26.

[232] *Ibid.*, par. 20.

[233] *Ibid.*, par. 26.

[234] *Ibid.*, par. 27.

[235] Especially to ESCR-Committee, *General Comment No. 15: The right to water (Art. 11 and 12)*, HRI/GEN/1/Rev.9 (Vol. I), 27 May 2008, at 97 (hereinafter "ESCR-Committee, *General Comment No. 15*"), par. 56, which *inter alia* reads: "Where such action is based on a person's failure to

Maybe the most important, and most recent, of case decisions on the right to water in South Africa is the case of *Mazibuko*. In this case,[236] the High Court of South Africa ruled that the City of Johannesburg's practice of forced installation of pre-paid water meters was unconstitutional. According to this practice, households were required to purchase pre-paid water cards in order to use water services; once the amount on a card was exhausted, a new card must be purchased. In essence, the meters function like prepaid phone cards.[237] The court directed the city instead to provide the residents of the poor neighborhood in question with the option of a "normal metered" water supply, as common in other (wealthier) parts of the city. Otherwise Johannesburg's water policy, to include pre-paid water meters in poor areas, was discriminatory. While people in low-income townships were required to pay for water in advance (and would not receive it if they were unable to do so), those in wealthy—and historically white parts—were entitled to water on credit, or to negotiate payment conditions with the provider for delayed payment of water bills. The installation would have similar effects to the effects of a disconnection: if a household is unable to pay for new cards, it would be cut off.[238]

Most remarkably, the High Court also ordered the City to provide residents of the poor neighborhood with 50 L of *free* water per person per day. This is an increase from the South African practice at the time, in particular in the city of Johannesburg, whereby each household (with an average household size of 16 persons) was provided with 200 L per day. Judge Moroa Tsoka, who delivered the judgment in *Mazibuko*, noted that 25 L per person would be insufficient, especially for those recipients suffering from special health conditions, in particular for people with the human immunodeficiency virus (HIV)/acquired immunodeficiency syndrome (AIDS).[239] The court also reminded the city that it received funds from the national government for water provision, and that it was the city's duty to use these funds for the benefit of the poorest of the poor.

The *Mazibuko*-case was appealed against in the High Court of South Africa where it was decided in March 2009.[240] The High Court of Appeals confirmed the findings of the first judgment with respect to the unlawful practices of the pre-paid

---

pay for water, their capacity to pay must be taken into account. Under no circumstances shall an individual be deprived of the minimum essential level of water."

[236] High Court (Witwatersrand Local Division) of South Africa, *Lindiwe Mazibuko and Others v. City of Johannesburg and Others*, Case No. 13865/06, Judgment of 30 April 2008, http://www.legalbrief.co.za/filemgmt_data/files/Mazibuko%20%26%204%20Others%20v%20The%20City%20of%20Johannesburg.pdf.

[237] Winkler (2010), p. 426.

[238] High Court (Witwatersrand Local Division) of South Africa, *Lindiwe Mazibuko and Others v. City of Johannesburg and Others*, Case No. 13865/06, Judgment of 30 April 2008, http://www.legalbrief.co.za/filemgmt_data/files/Mazibuko%20%26%204%20Others%20v%20The%20City%20of%20Johannesburg.pdf, par. 84.

[239] *Ibid.*, par. 179.

[240] Supreme Court of Appeal of South Africa, *The City of Johannesburg and Others v. Lindiwe Mazibuko and Others*, Case No. 489/08, Judgment of 25 March 2009, http://www.saflii.org/za/cases/ZASCA/2009/20.pdf.

water meters.[241] However, the court's reasoning was much more technocratic than the reasoning of the High Court in stating that "the City's Water Services by-laws do not authorize the installation".[242] The court gave the city 2 years to rectify the by-laws, but did not consider whether the manner in which the meters were generally installed was fair. With respect to the Free Water Basic Policy, the court reduced the suggested minimum of free water supply for the poorest of the poor, unable to pay for their water, from 50 to 42 L.[243] However, the court admonished the City of Johannesburg, and the local water supplier, to further develop policies to make the right to water, as laid down in section 27 of the South African constitution, as effective as possible.

The City of Johannesburg appealed against this decision to the Constitutional Court of South Africa, which delivered its final decision in October 2009.[244] The findings of the constitutional court were distinctly different from the one of the High Court and the High Court of Appeal. The Constitutional Court highlighted that section 27 of the constitution only entailed an obligation to take reasonable legislative and other measures to seek the progressive realization of the right to water over time.[245] The Free Basic Water policy, therefore, only needs to pass a test of "reasonableness",[246] which, in the eyes of the court, it did.[247] The court furthermore stated that it was inappropriate for a court to give a quantified answer to the question of what "sufficient water" was,[248] as had both of the other courts done.[249] The Constitutional Court called for the acceptance of the government's decision on this issue (25 L per person daily; or 6 kL per household monthly), as a decision of democratic value. On the issue of the pre-paid water meters, the court found that national legislation and the City's by-laws authorized the introduction of the meters (overruling the Court of Appeal), and that their introduction had not been unfair or discriminatory (overruling the High Court). Thus, the Constitutional Court fully upheld the appeal of the city of Johannesburg.

2.2.1.3 Analysis of the Protection of the Right to Water in South Africa

The protection and promotion of the right to water through South African laws, most noticeably through the constitution's section 27, is very sophisticated. Access

---

[241] *Ibid.*, par. 58.

[242] *Ibid.*, par. 58.

[243] *Ibid.*, par. 62.

[244] Constitutional Court of South Africa, *Lindiwe Mazibuko and Others v. City of Johannesburg and Others*, Case No. CCT 39/09, Judgment of 8 October 2009, http://www.saflii.org/za/cases/ZACC/2009/28.pdf.

[245] *Ibid.*, par. 58.

[246] *Ibid.*, par. 60.

[247] *Ibid.*, par. 143.

[248] *Ibid.*, par. 62.

[249] *Ibid.*, par. 68.

to water is phrased as an explicit fundamental constitutional right, which shares all the promises of other constitutionally guaranteed rights. The constitution does not distinguish between civil and political rights on the one hand, and economic, social and cultural rights on the other. They are all, in theory, equally enforceable[250]; the State is equally obliged to promote them, at least within its available resources. This general right to water laid down in section 27 of the South African constitution is specified by national laws. The National Water Act of 1998 on the one hand emphasizes the quality of water as a common public interest and its importance for social equity by establishing a progressive tariff-system. The Water Services Act of 1997 lists individual entitlements, establishing the unconditional right to have access to basic water services, and puts disconnections under very strict conditions. This strong social dimension of the right to water, evident in both laws, should be seen in connection with South African history until the mid 1990s. The principle of racial—and, as a corollary, social—equality was of high importance in the drafting phase of the new South African constitution in the 1990s and in the time after, when the two Water Acts were passed. In former days, access to water was largely dependent on land-ownership, from which black-Africans were widely excluded.

Turning to the judiciary, one has to acknowledge that the courts in early periods were reluctant in connecting the right with concrete duties for the State. As judge Niles Dunér stated in the *Manqele*-case, he did not find the right to have access to water as laid down in the Water Services Act guideline enough to ensure a basic minimum supply of water for everybody, irrespective of the ability to pay. His view was motivated by the assumption that judges have limited competence to deal with political matters. He assumed a narrow role for the judge. This decision, expressing the self-perception of a quite closed-off role of the judge, has been rightly criticized in the literature. This approach was essentially overruled in the *Bon Vista*-case where the court accepted exactly what the court in *Manqele* had rejected: the right to a basic water supply under all conditions as a crucial element of the right to water, most relevantly even without payment in cases of a proven inability to pay.

This dramatic shift in the judicial approach was caused by two main factors. Firstly, the applicants in the *Bon Vista*-case had not only relied on their right stemming from the Water Services Act, but also on the right to water enshrined in the constitution itself. The judge obviously had more respect for this constitutional right. It is unlikely that the judge in *Manqele* would have called the right to water a right "without content", if the right in question had been the fundamental right to water embedded in the constitution. Secondly, the court in *Bon Vista*, as it was interpreting a right of the South African Bill of Rights, also had to take international law into account. International law, in particular General Comment No. 15, already offered a more sophisticated view on the right's concrete shape and the accordant State duties.[251] Therefore, the judge in *Bon Vista* felt inclined to add

---

[250] Unknown (2007), p. 1067.

[251] See in particular ESCR-Committee, *General Comment No. 15* and discussion thereof in Sect. 2.3.2.2 of this book.

more weight to the right to water. Consequently the court prioritized it over the antagonistic interests of private water suppliers.

In the *Grootboom*-case, the court emphasized that the South African State is constitutionally bound not only to protect, but also to *fulfill* social and economic rights, and that victims of violations of these rights can rely on the courts for appropriate legal relief. It was very much a continuation of the jurisprudence advanced in the *Bon Vista*-case.

The recent *Mazibuko*-case might be the most remarkable of all the South African water cases. The original judgment not only promoted awareness for the social and economic context of the poorest of the poor. It also tried to uncover the inconsistency of constitutional guarantees and some policy practices with respect to the right to water. By looking into scientific studies, the judge in *Mazibuko* very neatly tried to make an argument for what the effective protection of the right to water for all must entail, if such a guarantee is taken seriously. The decision annulled the practice of the pre-paid meters, and at the same time increased the free amount for households to 50 L per person. The court thereby, willingly or not, could have created a precedent for litigation globally. However, the *Mazibuko*-case must be explicitly seen in the South African context, in which an outstanding legal guarantee of the right to water on the one hand, but an incomplete implementation on the other went sadly hand in hand.

However, even in the South African context, the original judgment of *Mazibuko* must be severely questioned. The main arguments of the appeal court were that the original decision infringed upon a competency that should be left to the legislative organs: to decide how much (free) water access is needed and can currently be afforded, in order to fulfill the right to water. However, the appeal was less successful than one might have expected: it confirmed the finding of the High Court, but simply mildly reduced the suggested basic free water supply. The Constitutional Court, however, tore those two decisions to pieces: not only did the court rule the pre-meter practice (and thereby possible disconnections) lawful, but also admonished the previous judges for their activism. They claimed a role, so the constitutional court, which was not theirs, but rather the role of the legislator and the administration. This was a slap in the face not only for the previous judges, but also for the implementation of the right to water through the South African courts. If the jurisprudence of the Constitutional Court is continued, this would mean a fall-back to the approach in *Manqele*, whereby the courts would accept a huge margin of appreciation for the legislator, and deny their role in "filling" gaps left by legislators. Although the original *Mazibuko*-decision invited accusations of judicial activism by doubling the offered amount of free water, the decision of the constitutional court reversed this trend too much, disclaiming almost all responsibility for the realization of the right, and leaving an almost free hand to the legislator and the administration.

Such an approach overlooks the fact that, despite the seemingly tight protection of the right in the law, severe problems with water supply in South Africa remain.

South Africa has witnessed severe water price increases over the last years,[252] as the water suppliers usually apply the principle of "cost-recovery": the assumption that consumers, in principle, are charged the full cost of the provision of water.[253] The Free Basic Water policy, generally a good policy, does not reach all poor households: according to the 2007/2008 annual report of the Department of Water Affairs and Forestry, more than a quarter of the poor households did not benefit from the Free Basic Water Policy.[254] But more recent numbers suggest a significant improvement on that area.[255] However, disconnections of water services have not been consistently ruled unlawful by the courts. At the same time, the installation or prepayment meters, which often have similar effects as cut-offs, has even been approved by the constitutional court in *Mazibuko*, even if only under strict conditions. Finally, much water infrastructure, necessary to ensure that water reaches all households, is still missing. According to the Department of Water Affairs and Forestry, in 2008, almost three million people were without access to improved drinking water.[256]

Despite these problems, the obligations to respect, protect, and fulfill the right to water are taken seriously in South Africa. One cannot apply the same standard as in some of the water-abundant developed nations with regard to water supply. But since the introduction of the fundamental right to water in the constitution, and its concretization in ordinary laws, the situation has much improved. This shows that not only in the richest countries of the world can the right to water be made effective. The hand-in-hand cooperation of both legislator and judges have made the legal protection and promotion of the right to water in South Africa unique amongst many States, despite the recent drawback of the decision of the constitutional court in *Mazibuko*. However, much work is still to be done: South Africa is far from having solved its water problem, in particular with respect to the poorest of the poor. One lesson to be learned is that creating a legal framework establishing a right to water alone is not enough in order to make the right effective.

The question remains: what is the right approach of South African judges towards the implementation of a right to water in the future? Courts should fill gaps where they were left open, thus where the law is not specific enough to put the right into practice, but not claim that there are gaps in the law, where there are none. There will be many unforeseen practices (like the one of the pre-paid water meters) in the future, as water suppliers have always proven to be creative in finding new ways of cutting down access of those customers that do not pay their bills. These are the cases where the judges must step in, and derive from general principles in the

---

[252] Partzsch (2006); Winkler (2010), p. 430, who specifically points out that prices have doubled in Johannesburg and tripled in Queenstown, Eastern Cape over the last years.

[253] McDonald (2002), p. 157.

[254] Department of Water Affairs and Forestry (2008), p. 58.

[255] See Department of Water Affairs (2012), p. 3.

[256] Department of Water Affairs and Forestry (2008), p. 55. See also the statistic for South Africa at http://www.unwater.org/statistics_KWIP.html.

law (e.g. the one on basic supply or pricing based on equity) concrete imperatives and prohibitions for the service providers. However, where the law has made these choices not *in abstracto*, but *in concreto*, there is much less room for judicial action. Here, indeed, the judicial task is restricted to control whether the practice allowed for in the law is generally reasonable and in accordance with higher sources of law, be they national or international.

2.2.2 *The Right to Water in Indian Law*

The general situation of water supply in India is curious. Although the country accounts for 4 % of the water resources of the entire world, water is a scarce resource in India.[257] Reasons for this scarcity is the steadily growing Indian population of more than one billion people,[258] and the fact that about 70 % of India's surface water resources, and an increasing number of India's groundwaters, are contaminated by biological, inorganic or toxic organic pollutants.[259] However, according to recent studies, 92 % of the Indian population has "access to improved water sources".[260] Still, 80 % of children suffer from water-borne diseases; an absolute number of more than 44 million people suffer from illnesses that relate to poor water quality.[261] All in all, according to the World Water Development Report 2003, India was nearly bottom of the entire table of all countries in terms of the quality of the available water: the 120th position out of 122 compared countries[262] is an alarming signal for the world's biggest democracy.

2.2.2.1 The Protection of the Right to Water through Indian Laws

India has neither a justiciable right to water in its constitution nor does any other Indian law establish an explicit right to water, although a variety of Indian laws deal with the topic of water in its different aspects; the *Indian River Boards*

---

[257] See the Draft National Water Policy (2012) of the governmental National Water Board, http://mowr.gov.in/writereaddata/linkimages/DraftNWP2012_English9353289094.pdf, par. 1.1. See also Muralidhar (2006), p. 65.

[258] See for instance India's Data Profile at the website of the World Bank Group, http://data.worldbank.org/country/india.

[259] Murty and Kumar (2011).

[260] See the United Nations (UN) Water Statistics for India, http://www.unwater.org/statistics_KWIP.html. Those numbers are an improvement to earlier numbers reported by Pant (2003), p. 16, and Bluemel (2004), p. 981, according to which 20 % of the population did not have access to water.

[261] Pant (2003), p. 15; Bluemel (2004), p. 981.

[262] UNESCO (2003), p. 140. See also Muralidhar (2006), p. 65.

Act,[263] the *Inter-State River Water Dispute Act*,[264] the *Water Act on Prevention and Control of Pollution*[265] and the *Environment Protection Act*[266] are the most important examples.[267] However, while most of these laws share the idea of the prevention of environmental pollution with regard to water-bodies, they do not establish rights for the individual, nor would they provide a suitable legal source for the derivation of such a right.

2.2.2.2 The Protection of the Right to Water through Indian Courts

In the absence of such an explicit right in the law, it was the Indian judiciary that have became active. The Indian judiciary derives the right to water from Art. 21 of the Indian constitution, which recognizes the right to life.[268]

The background of this legal procedure lies in the structure of the constitution itself. After its independence from Great Britain in 1947, India gave itself a, at the time, progressive constitution that recognized many of the fundamental human rights as laid down in the Universal Declaration of Human Rights (UDHR).[269] However, the recognized rights were divided into two categories: "Fundamental Rights" in the sense of civil and political rights on the one hand, and "Directive Principles of State Policy" covering the wide field of economic, social and cultural rights on the other. Only the category of civil and political rights is directly justiciable under the Indian constitution. By contrast, Art. 37 of the Indian Constitution makes Directive Principles of State Policy explicitly *non*-justiciable[270]: they are only principles to provide guidance to the legislator. If seen as a Directive Principle, or derived thereof, the right to water would thus not be enforceable *per se*.

---

[263] The River Boards Act, Act No. 49, 12 September 1956, http://www.indiankanoon.org/doc/1608688.

[264] The Inter-State River Water Disputes Act, Act No. 33, 28 August 1956, http://theindianlawyer.in/statutesnbareacts/acts/i90.html, amended by Act No. 14, 28 March 2002, http://www.indiankanoon.org/doc/1048477.

[265] The Water (Prevention and Control of Pollution) Act, Act No. 6, 23 March 1974, http://theindianlawyer.in/statutesnbareacts/acts/w6.html.

[266] The Environment (Protection) Act, Act No. 29, 23 March 1986, http://theindianlawyer.in/statutesnbareacts/acts/e18.html.

[267] Muralidhar (2006), pp. 67–68.

[268] Art. 21 Constitution of the Republic of India, http://lawmin.nic.in/olwing/coi/coi-english/Const.Pock%202Pg.Rom8Fsss(6).pdf: "No person shall be deprived of his life or personal liberty except according procedure established by law".

[269] Shah (1997), p. 1080.

[270] Art. 37 Constitution of the Republic of India, http://lawmin.nic.in/olwing/coi/coi-english/Const.Pock%202Pg.Rom8Fsss(6).pdf: "The provisions contained in this part shall not be enforceable by any court, but the principles therein laid down are nevertheless fundamental in the governance of the country and it shall be the duty of the State to apply these principles in making law."

Instead, the Indian courts chose an expansive interpretation of the scope and the content of the right to life, which itself as a Fundamental Right is legally enforceable under the Indian legal system. This approach has given the Indians courts the opportunity to provide justiciability to a wide range of economic, social and cultural rights,[271] including the right to water. The rights connected to the environment were also affirmed by the Indian Supreme Court on many occasions.[272]

In the groundbreaking decision *Francis Coralie Mullin*[273] of 1981, the Indian Supreme Court declared that the right to life includes the right to live with human dignity and "all that goes with it".[274] It admitted that the components of this right depend upon the extent of economic development of a country, but that the right, at least, must include the bare necessities of life, such as adequate nutrition, clothing and shelter.[275]

In 1988, the Supreme Court had to deal with the problem of pollution of the Ganges river by a municipal council. In the *M.C. Mehta*-case,[276] the court emphasized that due to the grave consequences of the pollution of water and air, the need for protecting and improving the natural environment was one of the fundamental duties under the Indian constitution.[277] Consequently, the court imposed the duty on the council to improve its sewerage system, and to ensure that throwing burnt and semi-burnt corpses into the river, as had been the practice in the Ganges region, was stopped.[278]

Specifically concerning a subjective right to water, the court stated in the case *Attakoya Thangal*[279] of 1990, which dealt with the legality of extracting water to augment water supply, that the right to drinking water and the right to fresh air were attributes of the right to life.[280] These were the basic elements, so the court, to sustain life itself. This ruling was confirmed throughout the 1990s in cases like

---

[271] See Powlowski (2006), p. 300; Unknown (2007), p. 1081; Muralidhar (2004), p. 25.

[272] See Allahabad High Court, *S.K. Garg v. State of U.P.*, 28 May 1998, (1988) 2 UPLBEC 1211, http://indiankanoon.org/doc/898522/; Supreme Court of India, *M.C. Mehta v. Union of India and others*, 12 January 1988, 1988 SCR (2) 530, http://indiankanoon.org/doc/59060; Supreme Court of India, *Subhash Kumar v. State of Bihar*, 9 January 1991, 1991 SCR (1) 5, http://indiankanoon.org/doc/1646284; High Court of Kerala, *Attakoya Thangal v. Union of India*, 1 January 1990, http://www.indiankanoon.org/doc/1980528; see for an overview of relevant case law Rosencranz and Rustomjee (1995), p. 328; Pant (2003), pp. 13 ff.; Unknown (2007), pp. 1080 ff.

[273] Supreme Court of India, *Francis Coralie Mullin v. The Administartor, Union Territory of Delhi*, 13 January 1981, 1981 SCR (2) 516, http://www.indiankanoon.org/doc/78536.

[274] *Ibid.*, par. 6.

[275] *Ibid.*, par. 6.

[276] Supreme Court of India, *M.C. Mehta v. Union of India and others*, 12 January 1988, 1988 SCR (2) 530, http://indiankanoon.org/doc/59060; for an in-depth discussion, see Rosencranz and Rustomjee (1995), p. 326.

[277] *Ibid.*, par. 29.

[278] *Ibid.*, par. 28.

[279] High Court of Kerala, *Attakoya Thangal v. Union of India*, 1 January 1990, http://www.indiankanoon.org/doc/1980528; also see Rosencranz and Rustomjee (1995), p. 327.

[280] *Ibid.*, par. 7.

Virendra Gaur,[281] in which the court found water pollution to be a possible violation of the right to life, and in *Vellore Citizens Welfare Reform*,[282] in which several tanneries were found in violation of the citizens' right to life by discharging untreated effluents into agricultural areas and local water supplies, and thereby severely polluting drinking water.

In the *FK Hussain*-case[283] the High Court of Kerala judged on a scheme of the administration to extract additional groundwater, thereby enabling saltwater to intrude the freshwater of some coral islands and making it unfit for human consumption. The court ruled against this practice, emphasizing that the right to freshwater and free air were basic elements of life itself, and considered them of attributes of the right to life of the Indian constitution.[284]

In more recent decisions, for instance, in the *Narmado Bachao Andolan-case*,[285] where the court had to deal with the decision of the Indian government to construct 3,000 dams on the Narmada river, the Supreme Court confirmed unambiguously that "water is the basic need for the survival of human beings and is part of the right to life and human rights as enshrined in Art. 21 of the Constitution of India". However, this time the court went even one step further and implied a positive obligation for the State, in saying that the right "can be served only by providing a source of water where there is none".[286] Provincial High Courts started to follow this general direction of the Supreme Court.[287]

2.2.2.3 Analysis of the Protection of the Right to Water in India

The protection and promotion of the right to water finds no direct legal basis in Indian law. Given a special distinction in the Indian constitution making only civil and political rights enforceable, the Indian High Courts have established the desired protection of the right wisely "through the backdoor": by choosing the right to life as its main legal anchor. This rights-based approach to the human need for water was, remarkably enough, developed by the Indian courts already in the early 1980s.

Given this situation the Indian courts found themselves in—their hands for the protection of social-economic rights being tied by Art. 37 of the Indian constitution—they did their best to give legal effect to the right to water. However,

---

[281] Supreme Court of India, *Virendra Gaur and Others v. State of Haryana and Others*, 24 November 1994, http://indiankanoon.org/doc/114802.

[282] Supreme Court of India, *Vellore Citizens Welfare Reform v. Union of India*, 28 August 1996, http://indiankanoon.org/doc/1934103.

[283] High Court of Kerala, *F.K. Hussain v. Union of India*, 26 February 1990, AIR 1990 Ker 321, http://www.indiankanoon.org/doc/1965629.

[284] *Ibid.*, par. 7.

[285] Supreme Court of India, *Narmada Bachao Andolan v. Union of India*, 18 October 2000, http://indiankanoon.org/doc/1938608.

[286] *Ibid.*, Conclusion of the case, p. 72 of pdf-version, http://indiankanoon.org/doc/1938608/.

[287] Bluemel (2004), p. 981; Pant (2003), pp. 13–16.

on closer inspection of the Indian cases, one has to concede that only few of the judgments make mention of a *positive* right to water or any *positive* governmental obligation to fulfill this right. The right to life as a civil and political right was for a long time seen as a predominantly *negative* right: this was the price to be paid for deriving it from the right to life (traditionally seen as a negative rather than a positive right). The courts may also not have wanted to overstrain the State and its financial capacities. However, in more recent judgments, as in the *Narmada Bachao Andolan*-case, the Indian courts have shown some more willingness to move into the direction of accepting the positive aspect of a right to water, including concrete entitlements for the individual. Where access to water is not yet existent, the State has a positive duty to develop this access in the future.

It was the negligence, or even the blindness, of the constitution towards water (e.g. no corresponding Directive Principle) that opened the door to the courts for choosing an approach linked to the right to life. If water protection had been mentioned as a Directive Principle in the constitution, it would have been hard, if not impossible, for the courts to link access to water to the right to life. Such an approach would have been seen as in conflict with the text of the constitution. A lesson learned in this respect is that a lack of regulation through the legislator cannot only be a problem, but also an opportunity. The courts could decide to give the right a much stronger anchor, linking it to the right to life, than the legislator ever would have done by making it a (non-enforceable) Directive Principle.

Despite this laudable judicial approach, India faces huge water problems. The second lesson the examples of India teaches is that a bare right on its own, which is not backed up by the political will or capacity of the State to respect, protect and fulfill it, is not worth much for the millions of people without access to water. Water law, water governance and the implementation of the right always have to go hand in hand.

Finally, the Indian example also shows how religious and cultural practices have to be taken into account, when assessing an appropriate protection of a right within a country. In Hinduism, water is not only seen as source of life, but as a purifying power for body and soul: ritual washings and ritual funerals in holy rivers, such as the Ganges are part of the religion. Legislators and judges in India will have to take these special conditions into consideration, as they did for instance in the *Mehta*-case. Strict international standards will only partly help to find adequate answers to balance these rituals with the demands of a right to water. The realization of the right must always be undertaken with due consideration for the specific national legal framework (e.g. the distinction between fundamental rights and fundamental principles) and cultural traditions (e.g. the special rituals of Hinduism that relate to water). With water being a crucial part of many cultural and religious practices in all different countries, such a cultural-relativist approach is even more important in the case of the right to water than it is for many other rights.

2.2.3 Comparative Analysis of the South African and Indian Approaches

Both the South African and the Indian models take a rights-based approach: on the one hand via an explicitly recognized constitutional right, on the other hand via a derivative right of the constitutional right to life. Thus, it is the individual that finds himself in the center of the legal focus in both cases.

In South Africa, this focus on the individual has to do with the country's bitter experiences of racial and social discrimination, which has motivated the constitutional legislator to emphasize not only civil and political, but also social and economic rights. They are expression of the hard-earned principle of social equity in the new South African constitution. In India, the rights-based approach is an expression of the alarming situation that the Indian population still finds itself in with respect to access to water of sufficient quantity and acceptable quality. It almost appears as if the courts wanted to "shake the legislator up", opposing unreliable State policies by reliable judicial means.

Although this humanistic approach unites the Indian and the South African approach, there are also significant differences. It is in one case the legislator supported by the courts, and in the other case the judiciary almost alone, that took responsibility for the protection of the right to water. This reiterates some of the problems discussed in the European examples, namely the legitimate role of the judge and the practical implementation of the rights in question.[288] A problem with the Indian approach in particular is that the right to life, in traditional understandings, has often been seen as a negative right. Therefore, the judges will in the future have to answer the question whether they accept the negative element of the right to water only, or are willing to go beyond.

If this was not the case, the Indian approach would be highly problematic: the very nature of the right to water lies also in the States' positive duty not only to respect, but also to protect and fulfill it. Otherwise, the right to water becomes indeed insipid, without much value and content of its own. A legal approach that purely bases the right to water on the right to life would therefore need to be "creative",[289] daring to stretch the right to water to encompass both a positive and a negative side, thereby leaving the constraints of the right to life's potentially negative nature behind. This "creativity", however, can also go too in extending legal protection, as seen in the original South African *Mazibuko*-case. States, especially the poorest of the poor, may be well advised therefore to set out positive minimum entitlements for the individual themselves before the judges do it—and potentially in a much more ambitious manner than the States would have wanted or can afford it. The South African model offers a well-balanced solution in this regard. The legislator keeps the decision within its own remit as to *which* rights

[288] See discussion above for the European examples, above at Sects. 2.1.1.1–2.1.1.3.

[289] Unknown (2007), p. 1086.

shall be legally protected. The judiciary reviews the "reasonableness", as Art. 27 par. 2 of the South African constitution puts it, of State action implementing this protection.

In sum, one must acknowledge that both chosen examples are rather positive ones with respect to the right to water: both legal systems generally recognize such an individual right. Most States have not gone so far as South Africa and India. The examples show two things: Firstly, a recognition of the right to water is not a legal idea that is only discussed in the ivory tower of academia and amongst representatives of water-abundant States. The right to water has started to become a component of various legal orders across the globe, even in countries where its recognition appears difficult. Secondly, the recognition of a right to water does not solve water problems in itself. The factual situation in India is still devastating as evidenced above, and that of South Africa is, despite noticeable progress, still problematic. However, both systems have given more or less reliable legal means to the individual: one of the first major steps on the way to a fair water distribution, within societies and amongst nations.

2.3 A Human Right to Water in International Law

Hardly any person, nor any State, refutes the normative claim that all humans *should* have a right to water of sufficient quantity and quality.[290] However, in order to positively assume the existence of such a right, it would need to meet certain criteria. It is not possible in a short chapter like this to assess comprehensively the define human rights, nor do I suggest that the criteria I will spell out in this chapter apply to all other human rights. However, if the term "human right to water" is to be taken serious, it must at least meet the three characteristics that inherent in the term already: it would need to be comprehensive, legally binding and (to a certain degree) self-standing. Firstly, as a "human" right, it would need to be generally applicable to all human beings, not only to particular groups or individuals.[291] I will call this the element of "comprehensiveness". As a "right", different from a mere political goal or a purely moral claim, it would have to be accepted by States as binding upon them to some degree. A right is more than a purely political recognition or a statement of intent. I will call this element "legal bindingness". Thirdly, is "water" a right of its own? Thus, we do not need a right publish articles as it is covered by the freedom of opinion or to own a car as it part of a larger and more general right to property. Only if the right has some distinct content, if it more than

[290] McCaffrey (2005), p. 93, even claims that "no one" disputes such a claim.

[291] I acknowledge that certain rights that are tailored for certain groups—rights of minorities or rights of women and children—and we still call them human rights. The idea here is that certain groups require specific protection. However, if we talk about a human right, we talk a right for all; access to water is necessary for a life in dignity of all human being, not matter their gender, ethnic origin, etc.

many claims under another right, we would think of it as a right of its own. I will call this the element of "self-standingness".

I introduce these characteristics purely for reasons of practicability: they do not lay claim to be exhaustive nor to be transferable to other rights. As we will see, they are tools to put the finger in the pitfalls that the "human right to water" has faced in international law for a long time.

2.3.1 Protection of the Right through International Conventions and Treaties

In order to assess whether such a right, meeting these criteria, is currently existent in international law, one firstly has to turn to the texts of international conventions and treaties,[292] as the first of the three sources of international law listed under Art. 38 of the ICJ Statute.

There are a number of international treaties that recognize such a right. Art. 14 (2)(h) of the *Convention on the Elimination of Discrimination against Women*[293] from 1979 obliges State parties, currently 185 in number, to ensure to rural women the right to enjoy adequate living conditions, particularly in relation to water supply. Art. 24(1) and (2)(c) of the *Convention on the Rights of the Child* (CRC),[294] one of the most widely ratified of all human rights treaties,[295] from 1989, commits State parties to implement children's rights to health by taking appropriate measures to combat disease and malnutrition through, *inter alia*, the provision of clean drinking water. In equivalent African human rights treaties on the same specific legal matters, such as the *African Charter on the Rights and Welfare of the Child* and the *Protocol to the African Charter on Human and Peoples' Rights on the Rights of Women in Africa*, there are similar references to the right to water.[296] Furthermore, International Humanitarian Law provides special protection during armed conflicts with respect to water. Geneva Conventions III and IV state that sufficient drinking water is to be supplied to prisoners and other

[292] This section will provide an overview only over the most crucial provisions, simply for reasons of space. For an exhaustive list, see COHRE (2008), p. 1.

[293] UNGA, *Convention on the Elimination of Discrimination against Women* (CEDAW), A/34/180, adopted on 18 December 1979, Annex, entered into force in 1981.

[294] CRC, 20 November 1989, 1577 UNTS, 3, entered into force in 1990.

[295] Only the United States of America (USA) and Somalia have so far not ratified the CRC.

[296] See Art. 14(2)(c) AfCRC, 11 July 1990, OAU, Doc. CAB/LEG/24.9/49 (1990), entered into force 1990, which demands that "State parties [...] shall take measures to ensure the provision of adequate standard of nutrition and safe drinking water" and Art. 15(a) Protocol to the African Charter on Human and Peoples' Rights on the Rights of Women in Africa, 11 July 2003, http://www.africa-union.org/root/au/Documents/Treaties/Text/Protocol%20on%20the%20Rights%20of%20Women.pdf, entered into force in 2005, according to which "State parties shall [...] provide women with access to clean drinking water [...]".

detainees, and that they are to be provided with shower and bath facilities.[297] The additional protocol to the Geneva Conventions of 1977 prohibits the destruction of "objects indispensable to the survival of the civilian population such as [...] drinking water installations and supplies and irrigation works."[298]

Along the three aforementioned parameters, all these recognitions of the right to water in international treaties are legally binding: States have ratified these treaties and accepted them as legally binding upon them. These recognitions are also self-standing: a right to water, or aspects of it, are not only implied, but explicitly recognized in these provisions. However, all these treaties are limited in which persons they provide with the right. The holder of the mentioned rights are only women, only children, only detainees, etc. None of these recognitions are comprehensive in nature. They do not apply equally to all human beings. None of them, by themselves, can thus be regarded as sufficient recognition of a human right to water. Such a right would by definition have to be applicable to all humans.

Secondly, there are numerous international declarations and resolutions on a human right to water. Already in the 1970s, the *Mar del Plata Declaration* of the 1977 UN Water Conference states that all people have the right to have access to drinking water in quantities and quality equal to their basic needs.[299] In the early 1990s, the *UN Conference on Environment and Development* recognized in its Agenda 21 that all people, whatever their stage of development and their social and economic conditions, have the right to have access to drinking water in quantities and of a quality equal to their basic needs.[300] Two years later, the Programme of Action of the *International Conference on Population and Development* in Cairo, attended by almost 180 States, stated that human beings have the right to an adequate standard of living, including adequate water and sanitation.[301] In the rise of the new millennium, the UN General Assembly (UNGA) reaffirmed in its resolution 54/175 on the right to development the rights to food and clean water as fundamental human rights.[302] In 2010, the UNGA released resolution 64/292; a resolution entirely dedicated to "The human right to water and sanitation", therein

---

[297] See Art. 20, 26, 29 and 46 Geneva Convention (III) Relative to the Treatment of Prisoners of War (1949), UNTS 75 (1950), 135; Art. 85, 89 and 127 Geneva Convention (IV) Relative to the Protection of Civilian Persons in Time of War (1949), UNTS 75 (1950), 287.

[298] Art. 54 Additional Protocol (I) to the Geneva Conventions relating to the Protection of Victims of International Armed Conflicts (1977), UNTS 1125 (1979), 3; Art. 5 and 14 Additional Protocol (II) Relating to the Protection of Victims of Non-International Armed Conflicts (1977), UNTS 1125 (1979), 609.

[299] Report of the UN Water Conference, Mar del Plata, 14–25 March 1977, E/CONF.70/29, 1977, pp. 66–67.

[300] UN Conference on Environment and Development, Rio de Janeiro, 3–14 June 1992, *Agenda 21*, http://www.unep.org/Documents.Multilingual/Default.asp?documentid=52.

[301] Report of the International Conference on Population and Development, Cairo, 5–13 September 1994, A/CONF.171/13, 18 October 1994, Principle 2.

[302] UNGA, *The right to development*, A/RES/54/175, 15 February 2000, par. 12.

recognizing the right to safe and clean drinking water and sanitation as a human right.[303]

All these declarations have, amongst others,[304] recognized or called for the recognition of a right to water. Many of them have outstanding political value. They have suggested recognition of the right in an explicit way by singling it out, thereby identifying it as a "self-standing" entitlement. They have also assumed the right to be a right for all human beings, regardless of gender, race or socio-economic status, making it a comprehensive right. However, they all share a flaw under international law: they are in nature not directly legally binding upon States. They are political affirmations, declarations of intent, global appeals—but they do not have the status of international hard law. This is not to downplay these recognitions. Many of them are mayor political achievements. However, they must be considered in themselves unsuitable to be the sole legal source for a right to water.

A special case is the most recent resolution 15/9 of the Human Rights Council (HRC) on "Human rights and access to safe drinking water and sanitation", which was adopted without a vote,[305] certainly an extraordinary achievement on the international agenda. The resolution affirms in its operative paragraph 3 that the human right to safe drinking water and sanitation is derived from the right to an adequate standard of living (and related to several other rights). This has led the Special Rapporteur to release a press statement that "this means that for the UN, the right to water and sanitation, is contained in existing human rights treaties and is therefore legally binding".[306] On web pages within the UN system[307] or outside,[308] this statement was simply copied and re-published. However, the statement can be easily misunderstood; the Special Rapporteur's words must be treated with caution. It is well established in international law that the resolutions of the UN General

---

[303] UNGA, *The human right to water and sanitation*, A/RES/64/292, 3 August 2010, par. 1.

[304] Again, space constraints do not allow an exhaustive discussion; it would also not add much value for the purpose of the research question. For a detailed discussion, see COHRE (2008), p. 1.

[305] This being a mayor and somehow unexpected achievement, one must acknowledge that the delegate of the United Kingdom in the HRC session explained that his country would not call for a vote, but disassociate his country from the resolution. However, the main reason for this behaviour was the UK's disagreement of the right to sanitation, which lies outside the scope of this study. The transcripts of the session not being published yet, the statement is available at the HRC's archived web stream, http://www.un.org/webcast/unhrc/archive.asp?go=100930.

[306] See UN Office of the High Commissioner for Human Rights, *UN united to make the right to water and sanitation legally binding*, 1 October 2010, http://www.ohchr.org/EN/NewsEvents/Pages/DisplayNews.aspx?NewsID=10403&LangID=E.

[307] Just to name one, see for instance UN News Centre, *Right to water and sanitation is legally binding, affirms key UN body*, 1 October 2010, http://www.un.org/apps/news/story.asp?NewsID=36308.

[308] See for instance Amnesty International, *UN reaffirms legal rights to water and sanitation*, 2 October 2010, http://www.amnesty.org.au/news/comments/23808; or Mistry (2010).

Assembly are not legally binding: Art. 10 and 14 of the UN Charter refer to UNGA resolutions as "recommendations", so that the recommendatory nature of UNGA resolutions is without doubt. Resolutions of the HRC, which was founded as a subsidiary organ of the UNGA,[309] undoubtedly share the same legal destiny: they are also "recommendations".[310] The human right to water could therefore not become legally binding as treaty law by the means of a single HRC resolution. This also flows from the HRC's composition: while in the UNGA representatives of all States are present and have a vote, the HRC, according to its institutional setup, comprises of 47 members only.[311] Accepting HRC resolutions as directly legally binding could therefore mean to coerce States to accept resolutions as legally binding upon them for which (or against which) they did not have the opportunity to vote. Such an outcome is most certainly incompatible with some of the most essential principles of international law, most particularly with the principle of "free consent", as for instance expressed in the preamble of the Vienna Convention on the Law of Treaties (VCLT)[312] and the requirement of recognition of international treaties by States before they have to accept them as source of law as expressed in Art. 38(1)(a) of the ICJ Statute.[313]

The HRC resolution did thus not directly change the legal status of the right to water as part of treaty law,[314] neither by creating a new right nor by providing a legally binding interpretation of Art. 11 ICESCR. However, on close inspection of the Special Rapporteur's words—which the cited media have not sufficiently undertaken—the expert pointed out that the HRC resolution "means that *for the UN*, the right to water and sanitation, is contained in existing human rights treaties and is therefore legally binding" (italics added). Thus, she did not claim the right was now legally binding upon States, but that *the UN* would accept it as part of Art. 11 ICESCR. Although there is no direct legal implication for this either, the UN itself might, given the HRC resolution being adopted without a vote, in the future not contest the existence of a right to water. Any other behavior might be considered as a case of "*venire contra factum proprium*" or "*estoppel*"[315]: a behavior

[309] UNGA, *Human Rights Council*, A/RES/60/251, 3 April 2006, op. par. 1.

[310] *Ibid.*, see the unambiguous language of the resolution at op. par. 3, 5c and 5i. The HRC recognizes this itself: on its webpage, it calls its main purpose "addressing situations of human rights violations and make *recommendations* on them" (italics added).

[311] *Ibid.*, op. par. 7.

[312] VCLT, 23 May 1969, UNTS 1155 (1980), 331, Preamble, par. 3: "*Noting* that the principles of free consent and of good faith and the *pacta sunt servanda* rule are universally recognized" (italics added).

[313] Art. 38 ICJ Statute, 26 June 1945, recognizes international treaties as sources of international law only insofar they "are expressly recognized by the contesting parties".

[314] Another question is what meaning such a resolution might have for international custom, see Sect. 2.3.3 of this book.

[315] On this principle, see in general: MacGibbon (1958), p. 468; Bowett (1957), p. 176; Müller and Cottier (1995), p. 116.

inconsistent with previous behavior and a violation of good faith,[316] which the UN most likely would aim to avoid.

In sum, neither the HRC resolution of 2010 nor any of the other statements political examined here provide a legal recognition of the human right to water that would be legally binding for States. Another question is, of course, the meaning of these documents for the right to water in international customary law.[317]

Finally, there are numerous other recognized human rights in international law that are closely related to the issue of water. Amongst them are the right to life,[318] the right to the highest attainable standard of physical and mental health[319] or an adequate standard of living including the right to housing or food.[320] Undoubtedly, these human rights have strong links to the basic need for water. If a human right to water were to be accepted, it would be tightly interwoven with these rights. However, in none of the binding human rights instruments applicable to all human beings—in particular neither the International Covenant on Civil and Political Rights (ICCPR) nor in the International Covenant on Economic, Social and Cultural Rights (ICESCR)—is there an explicit mention of an explicit human right to water, not to speak of a definition of its normative content.

In conclusion, if one were to look for a human right to water in international treaties, it is not easily found. Testing the right to water along the criteria that I have suggested (with the caveats that I have also pointed out)—their applicability for all humans ("comprehensiveness"), their design not only as moral but also legal claims ("legal bindingness") and the articulation as an explicit and spelled-out right ("selfstandingness"), the right would, however, fall short of meeting these criteria.

This might be a surprising finding at first. However, it is not the end of the story. If one re-conceptualizes one's expectations of the term "human right to water"—not expecting to find all three mentioned characteristics in one body, but rather accepting the right as a complex of different rights—one might still conclude that there is such a thing as a human right to water in international law[321]; it just

[316] The ICJ derives the principle of "estoppel" from the concept of good faith and equity, see for instance in ICJ, *Delimitation of the Maritime Boundary in the Gulf of Maine Area* (Canada v. USA), Judgment of 12 October 1984, ICJ Reports (1984), 246, 305.

[317] See Sect. 2.3.3 of this book.

[318] Art. 6(1) ICCPR (1966), UNTS 999 (1976), 171: "Every human being has the inherent right to life. This right shall be protected by law. No one shall be arbitrarily deprived of his life."

[319] Art. 12(1) ICESCR (1966), UNTS 993 (1976), 3: "The States Parties to the present Covenant recognize the right of everyone to the enjoyment of the highest attainable standard of physical and mental health."

[320] *Ibid.*, Art. 11(1), s. 1: "The States Parties to the present Covenant recognize the right of everyone to an adequate standard of living for himself and for his family, including adequate food, clothing and housing, and to the continuous improvement of living conditions. The States Parties will take appropriate steps to ensure the realization of this right, recognizing to this effect the essential importance of international cooperation based on free consent."

[321] On such a re-conceptualization, see Sect. 3.4.2 of this book.

requires creativity and flexibility in understanding how a human right can be composed in international law.

2.3.2 *Interpretations of Treaties through Courts and Quasi-Courts*

As we saw before in some of the national examples, judges (and quasi-judges such as arbitrators or experts) have been often very proactive for the promotion of a right to water. Similarly, international judges and quasi-judges have taken creative and innovative approaches to link the right to water to a variety of other accepted human rights. The recognition of a human right to water has therefore been termed an "emerging trend"[322] in the jurisdiction of international courts, quasi-courts and tribunals.

2.3.2.1 General Decisions of International Courts and Quasi-Courts

Already in the early 1980s in the *Port Hope*-case, where dumping of nuclear wastes within Port Hope, Ontario, was causing large-scale pollution of residences, the UN Human Rights Committee observed that the case raised serious issues under the right to life under Art. 6(1) ICCPR and emphasized therefore the State's water-related obligation to protect human life.[323] In the early 1990s, the African Commission interlinked in *Free Legal Assistance Group and Others v. Zaire*[324] water issues on several occasions with the right to health as laid down in Art. 16 of the African Charter on Human and Peoples' Rights. It stated that the failure of the government to provide basic services, such as electricity and drinking water,[325] constituted a "serious or massive" violation[326] of this article, which states that every individual shall have the right to enjoy the best attainable state of physical and mental health. In accordance with this article, State parties should take the necessary measures to protect the health of their people.

In the late 1990s, the International Court of Justice recognized in its *Advisory Opinion on the Legality of the Use of Nuclear Weapons*[327] the close connection

---

[322] Scanlon et al. (2004), p. 10.

[323] CCPR-Committee, *E.H.P. v. Canada*, Communication No. 67/1980, CCPR/C/17/D/67/1980, 27 October 1982, p. 20.

[324] African Commission on Human and Peoples' Rights, *Free Legal Assistance Group and Others v. Zaire*, Communication No. 25/89, 47/90, 56/91, 100/93, October 1995, http://www.achpr.org/files/sessions/18th/comunications/25.89-47.90-56.91-100.93/achpr18_25.89_47.90_56.91_100.93_eng.pdf.

[325] *Ibid.*, par. 47.

[326] *Ibid.*, final finding, after par. 48.

[327] ICJ, *Legality of the Threat or Use of Nuclear Weapons*, Advisory Opinion of 8 July 1996, ICJ Reports (1996), 226.

between environmental goods, of which water is certainly one, and the right to life. It stated that the environment represented the quality of life and the very health of human beings.[328]

In 1997, in the famous *Gabcíkovo-Nagymaros*-case[329] the International Court of Justice accepted the concerns expressed by Hungary for its natural environment such as groundwater and the water supply of Budapest related to an "essential interest of that State".[330] However, the Court recognized water explicitly only in general terms and in connection with a State's interest, not specifically with rights of individuals. Judge Weeramantry wrote in his separate opinion on the *Gabcíkovo-Nagymaros*-case that "the protection of the environment is [...] a vital part of contemporary human rights doctrine, for it is a sine qua non for numerous human rights such as the right to health and the right to life itself".[331] He continued to state that "damage to the environment can impair and undermine all the human rights spoken of in the Universal Declaration and other human rights instruments".[332]

In the same year, the Inter-American Commission on Human Rights found in its *Report on the Human Rights Situation in Ecuador*[333] that inhabitants exposed to toxic by-products of oil exploitation in their drinking and bathing water were subject to serious dangers to their lives and their health. It was argued that, where environmental contamination and degradation pose a persistent threat to human life and health, the rights to life, physical security and integrity were implicated.[334]

In the recent case *Pulp Mills* between Argentina and Uruguay,[335] in which both States argued over the use of the Uruguay River, the ICJ referred again to the rights of riparian States rather than individuals.[336] The court, however, argued that "the environment is not an abstraction, but represents the living space, the quality of life

---

[328] *Ibid.*, p. 241, par. 29.

[329] ICJ, *Gabčíkovo-Nagymaros Project (Hungary v. Slovakia)*, Judgment of 25 September 1997, ICJ Reports (1997), 7.

[330] *Ibid.*, p. 27, par. 53; see also Dupuy (2006), p. 1.

[331] ICJ, *Gabčíkovo-Nagymaros Project* (Hungary v. Slovakia), Judgment of 25 September 1997, Separate Opinion of Vice-President Christopher Gregory Weeramantry, ICJ Reports (1997), 88, 91.

[332] *Ibid.*, p. 91.

[333] Inter-American Commission on Human Rights, R*eport on the Situation of Human Rights in Ecuador*, OEA/Ser.L/V/II.96, Doc. 10 rev. 1, 24 April 1997, http://www.cidh.oas.org/countryrep/ecuador-eng/index%20-%20ecuador.htm.

[334] *Ibid.*, chapter 5; on this case, also see Shelton (2002), p. 158.

[335] ICJ, *Case concerning Pulp Mills on the River Uruguay* (Argentina v. Uruguay), Order on the Request for the Indication of Provisional Measures of 13 July 2006, ICJ Reports (2006), 113, 113; ICJ, *Case concerning Pulp Mills on the River Uruguay* (Argentina v. Uruguay), Judgment of 20 April 2010, ICJ Reports (2010), 14.

[336] ICJ, *Case concerning Pulp Mills on the River Uruguay* (Argentina v. Uruguay), Order on the Request for the Indication of Provisional Measures of 13 July 2006, ICJ Reports (2006), 113, par. 80: "to safeguard the continued conservation of the river environment and the *rights* of economic development *of the riparian States*" (italics added); *Judgment*, par. 177: "to reconcile the varied *interests of riparian States* in a trans-boundary context and in particular in the use of a shared natural resource" (italics added).

and the very health of human beings".[337] Given the court also emphasized the need to protect the environment and the water of the River Uruguay,[338] the judges suggested a subtle link between human rights interests and the quality of the water of the Uruguay River. However, the court made only this indirect, no direct reference to a human right to water. This is somewhat surprising given that Uruguay itself had affirmed its commitment to "respect [...] the environment and [...] the entire range of human rights of the Uruguayan and Argentine peoples".[339] It is noteworthy that the court had rendered its decision in early 2010, while the political affirmations of the human right to water at the UNGA and HRC occurred a few months later, namely in July and September of that year.[340] The court might have made mention of the human right to water, if the judgment had been rendered a few months later, namely *after* the two UN resolutions of 2010. It will be interesting to note how the court will decide in the case of *Construction of a Road in Costa Rica Along the San Juan River* between Nicaragua and Costa Rica[341]: will the judges finally make explicit mention of a human right to water?

2.3.2.2 General Comment No. 15

Most importantly, in 2002 the United Nations Committee on Economic, Social and Cultural Rights (ESCR-Committee), the body responsible for the oversight of the covenant of the same name, adopted its *General Comment No. 15 on "The Right to Water"*.[342] As the General Comment is by far the most relevant of all judicial or quasi-judicial recognitions of the right to water, and the one which elaborates the most on the normative content of such a right, it must be addressed with particular accuracy and diligence.

Genesis and Background of General Comment No. 15

Before the Committee even started its work on General Comment No. 15, it could draw upon its own practice: its concluding observations on State reports where water had become relevant[343] and its previous General Comments that partly dealt

---

[337] *Ibid.*, par. 72.

[338] *Ibid.*

[339] *Ibid.*, par. 56.

[340] UNGA, *The human right to water and sanitation*, A/RES/64/292, 3 August 2010; HRC Res. 15/9, "Human rights and access to safe drinking water and sanitation", 30 September 2010, UN Doc. A/HRC/Res/15/9 (2010).

[341] The status of the proceedings can be seen at the website of the ICJ, http://www.icj-cij.org.

[342] ESCR-Committee, *General Comment No. 15*.

[343] According to Riedel (2006), p. 19, before General Comment No. 15, in 33 out of 114 concluding observations, water rights were addressed and discussed.

with the issue of water, for instance General Comments No. 4 on housing,[344] No. 12 on food,[345] and No. 14 on health.[346]

Subsequent General Comments made reference to the right to water linking it with further topics, e.g. General Comment No. 20 on non-discrimination[347] or General Comment No. 21 on the right to take part in cultural life.[348]

The Committee made an effort to keep its language in General Comment No. 15 in line with previous General Comments, and focused on already developed descriptive and normative categories, such as *availability*, *affordability*, *accessibility* of a right. The *mis en oeuvre* of the General Comment is also feasibly influenced by the United Nations Millennium Declaration, in particular by the goal to halve the population without access to a basic water supply by 2015.[349]

Sources and Content of the General Comment

The General Comment starts by declaring water a limited natural resource and at the same time a public good fundamental for health and life, indispensable for leading a life with human dignity.[350] Like an overarching summary of the entire General Comment, par. 2 s. 1 states that "*[t]he human right to water entitles everyone to sufficient, safe, acceptable, physically accessible and affordable for personal and*

[344] ESCR-Committee, *General Comment No. 4: The right to adequate housing (Art. 11 par. 1)*, HRI/GEN/1/Rev.9 (Vol. I), 27 May 2008, at 11, par. 8(b) stating that "all beneficiaries of the right to adequate housing should have sustainable access to natural and common resources, safe drinking water [. . .]."

[345] ESCR-Committee, *General Comment No. 12: The right to adequate food (Art. 11)*, HRI/GEN/1/Rev.9 (Vol. I), 27 May 2008, at 55, which, however, interestingly enough does not mention the issue of water or liquid food.

[346] ESCR-Committee, *General Comment No. 14: The right to the highest attainable standard of health (Art. 12)*, HRI/GEN/1/Rev.9 (Vol. I), 27 May 2008, at 78 (hereinafter "ESCR-Committee, *General Comment No. 14*"), which defines water in several paragraphs as an underlying determinant of the right to health, see par. 4, 11, 12a, 15, 34 and 51.

[347] ESCR-Committee, *General Comment No. 20: Non-discrimination in economic, social and cultural rights (Art. 2 par. 2)*, E/C.12/GC/20, 2 July 2009 (hereinafter "ESCR-Committee, *General Comment No. 20*") which emphasizes, *inter alia*, in par. 8 that all individuals must have equal access to adequate water and sanitation to overcome discrimination against women and girls and persons living in informal settlements and rural areas.

[348] ESCR-Committee, *General Comment No. 21: Right of everyone to take part in cultural life (Art. 15 par. 1(a))*, E/C.12/GC/21, 21 December 2009, which stresses, *inter alia*, in par. 16 e) the need to take into account cultural values attached to the use of water.

[349] UNGA, *United Nations Millennium Declaration*, A/RES/55/2, 18 September 2000, par. 19: "*We resolve further [. . .] by the year 2015 [. . .] to halve the proportion of people who are unable to reach or to afford safe drinking water.*"

[350] ESCR-Committee, *General Comment No. 15*, par. 1.

domestic uses."[351] As for any human right, States Parties have to respect, protect and fulfill the right for water.[352]

After these introductory explanations, the Committee used Art. 11(1) ICESCR, dealing with the right of everyone to an adequate standard of living, including adequate food, clothing and housing, as a starting source to derive an independent right to water for personal and domestic uses. As the listed rights in Art. 11 (food, clothing, housing) were not exhaustive due to the use of the word "including", the Committee found that the right to water clearly fell into the category of those guarantees essential for securing an adequate standard of living.[353] The Committee derived this stand-alone right to water furthermore from the health provisions of the ICESCR. Art. 12(1) provides everyone with the right to the highest-attainable standard of health. In General Comment No. 14, the Committee had already stated that the "underlying components" of the right to health include potable water.[354] In General Comment No. 15, the Committee went further and elaborated a number of aspects relating to water under the right to health. It concluded that health could not be ensured effectively under unsafe or toxic water conditions.[355] The Committee also related the right to water to other rights enshrined in the Universal Declaration of Human Rights—in particular to the right to life and the notion of human dignity—[356] albeit to a much lesser degree than it did with respect to an adequate standard of living and health.

The General Comment continues to outline the normative content of the right to water. While accepting that different circumstances and conditions might influence the concrete shape and dimension of water access, the Committee claimed that everybody, on the basis of non-discrimination, has the right to sufficient, safe, accessible and affordable water of an acceptable quality.[357] Thus, water must be present in continuous *sufficiency* for personal and domestic use, in particular for the purpose of drinking, sanitation, washing of clothes, food preparation, and personal and household hygiene. Water *quality* refers to the fact that water must be free of microorganisms, chemical substances and radiological hazards so that water quality does not jeopardize a person's health. Water *accessibility* demands that water must be within physical reach. Water *safety* means that physical security, in particular for women who in many communities are responsible for fetching water, is ensured while accessing water sources. Finally, water *affordability* does not proscribe that water must be provided for free, but that the costs of accessing water must be realizable for all, including vulnerable and marginalized groups. Altogether, States

[351] *Ibid.*, par. 2.

[352] *Ibid.*, par. 20. See generally on the obligations of State parties, OHCHR (2005), pp. 15 ff.; Leckie and Gallagher (2006), pp. xx ff.; Riedel (2011), par. 14–15.

[353] ESCR-Committee, *General Comment No. 15*, par. 3.

[354] ESCR-Committee, *General Comment No. 14*, par. 11, 12a, 15, 34 and 51.

[355] ESCR-Committee, *General Comment No. 15*, par. 8.

[356] *Ibid.*, in particular par. 1, 3, 11 and 29.

[357] *Ibid.*, definitions par. 12 (a–c).

are called upon to respect, protect and fulfill the right to water[358]; a term that is not specific for the right to water, but has commonly been used to describe State obligations with respect to economic and social rights.

General Comment also outlines a few core obligations[359] that States must at the very least ensure and which are of immediate effect. These obligations include, according to the Committee, non-discriminatory access to the minimum essential amount of water for personal and domestic use, accessibility in reachable distances and without significant waiting times, and equitable distribution of water within the limits of availability.[360] To meet these obligations, the adoption of national water strategies and action plans, the creation of water access monitoring systems, programs to help the most vulnerable groups, and measures necessary to minimize water-related diseases, are requested.[361]

Meaning and Critique of the General Comment

General Comment No. 15 was certainly a milestone on the way towards the recognition of the human right to water. It is the most explicit recognition of an independent and generally applicable human right to water. It elaborates with great length on the normative content of the right, and introduces an approach to the sources of the right and its connection with other rights. It has been referred to by following General Comments[362] and judicial decisions on the national level.[363] It must be considered to be one of the most influential of all General Comments of the Committee.

However, there are also reasons for critique. Formulating critique is not downplaying the significance and success story of General Comment No. 15; it makes its appraisal even more credible than uncritical flattery would. The critique can be attributed to two spheres: General Comments as tools for international law-making in general, and the specific content and style of General Comment No. 15.

On the first level, one has to keep in mind that General Comments are only an authoritative, not a legally binding interpretation of the ICESCR.[364] General Comments are not part of international "hard" law. They are interpretations of an

---

[358] *Ibid.*, par. 20–29.

[359] ESCR-Committee, *General Comment No. 3: The nature of States parties' obligations (Art. 2 par. 1)*, HRI/GEN/1/Rev.9 (Vol. I), 27 May 2008, at 7, par. 10; very enlightening on this topic also: Young (2008), p. 113.

[360] ESCR-Committee, *General Comment No. 15*, par. 37.

[361] *Ibid.*, par. 37 (f–i).

[362] ESCR-Committee, *General Comment No. 19: The right to social security (Art. 9)*, HRI/GEN/1/Rev.9 (Vol. I), 27 May 2008, at 152, par. 8 and 28, footnotes; ESCR-Committee, *General Comment No. 20*, par. 6.

[363] See for instance the *Mazibuko* case, par. 34, discussed above in Sect. 2.2.1.2 of this book.

[364] See ECOSOC, *International Covenant on Economic, Social and Cultural Rights*, E/RES/1987/5, adopted on 26 May 1987, containing no authorization to release legally binding comments; also see Filmer-Wilson (2005), p. 213.

expert committee. This committee is, however, the only one of its kind to express general interpretations of the covenant; thus, the interpretations are highly valuable and influential. At the same time, they remain interpretations. They do not have the legal force to create new direct legal obligations for States. In this respect, the General Comment, like some of the other documents discussed before,[365] fails to meet the criterion of legal bindingness. The General Comment's importance is clearly to be found in the two other before mentioned categories: it establishes normative content and makes describes the contours of a self-standing right[366] and argues that the right should belong not to certain groups, but to all human beings.

On the level of the normative content of General Comment No. 15, Stephen Tully has critiqued General Comment No. 15 categorically,[367] challenged by two responses by Malcolm Langford.[368] Tully challenges the content, style and drafting process of General Comment No. 15. He comes to the conclusion, contrary to the General Comment's findings, that an entitlement to access water for personal or domestic use available to all does not exist under contemporary international law, and that the Committee should have been "more careful" in drafting the General Comment.[369]

Creation of Ever New Human Rights of Unclear Shape and Content?

Firstly, Tully critiques the creation of ever new human rights. This essentially goes back to the idea expressed by Philip Alston some 25 years ago that there is no need to conjure up ever new human rights for all and everything at the international level.[370] According to Tully, the term "adequate standard of living *including* adequate food, clothing and housing" (italics added) used in Art. 11 of the ICESCR should not be used as an invitation to create ever new rights related to an adequate standard of living. Otherwise, rights such as "postal delivery" and "access to the internet" would soon also be considered new human rights derived from Art. 11 ICESCR.[371] Even more, according to Tully, a right to water would (due to its interconnectedness with other rights) be broader than the right of an adequate living standard[372]; yet, for a derivative right to be broader than the right it is derived from would logically be incoherent. Altogether, deriving a right from a right blurry in shape and content, like the right to an adequate standard of living, would only create a weak right to water.[373]

---

365 See discussion at Sect. 2.3.1 of this book.

366 Riedel (2006), p. 19, uses the similar language of a "free-standing" right to water.

367 Tully (2005), p. 35; *id.* (2006), p. 461.

368 Langford (2006a), p. 433; *id.* (2006b), p. 473.

369 Tully (2005), p. 35.

370 Alston (1984), p. 607.

371 Tully (2005), p. 35.

372 Tully (2006), p. 461.

373 *Ibid.*

The first part of the argument—that including a right to water would mean to open up the gates for the inclusion of obscure rights, such as postal delivery or internet services—is not very convincing. Hardly any right could claim more vital importance than the right to water, which is also recognized by General Comment No. 15 when it states that "the right to water clearly falls within the category of guarantees essential for securing an adequate standard of living, particularly since it is one of the most fundamental conditions for survival."[374] Clearly, the Committee would not have stated something similar about a right to postal delivery. It will most likely only consider rights for further "inclusion" that are of vital comparable importance to those expressly mentioned. Tully therefore himself "conjures up" a scenario that is unrealistic.

However, Tully's thoughts on the problem of legal derivation are very interesting.[375] It seems only logical that a derivative right carries the characteristics of the right it is derived from. Assuming that socio-economic and civil and political rights are two different categories of rights, although without inherent hierarchy, it is plausible to derive socio-economic rights only from other socio-economic rights and civil and political rights only from other civil and political rights. This, in turn, means that a human right to water derived from an adequate standard of living must also be, in its entirety, a socio-economic right. A derived right cannot be stronger, or larger in scope, than its parent right. Tully suggests that the Committee, by drawing at the same time on the right to health and the right to an adequate standard of living, created a derivative right to water that is larger than each of its parents.

However, this is where Tully's argument loses coherence: a right which is derived from several rights *can* be larger than each of its parents: it combines different elements of those rights, and this new conglomerate can, of course, exceed each of its parents in broadness or even acceptance. It can just not exceed the sum of both of its parents' scope. Thus, the right to an adequate standard of living, read together with the right to health, was a suitable parent for the right to water with the suggested normative content. Another question, of course, is whether it was the optimal one to choose.

Excess of the Committee's Power in Contrast to the Drafters' Will?

Secondly, Tully claims that the Committee exceeded its interpretative competence through the General Comment and engaged in activism.[376] An amendment of the ICESCR following the complicated procedure of Art. 29 ICESCR[377] would have

---

[374] ESCR-Committee, *General Comment No. 15*, par. 3.

[375] See also the chapter on legal derivation in Sect. 3.4.1 of this book.

[376] Tully (2005), p. 35, referring to Dennis and Stewart (2004), p. 462.

[377] Art. 29(1) s. 1 ICESCR (1966), UNTS 993 (1976), 3 "Any State Party [. . .] may propose an amendment and file it with the Secretary-General of the United Nations. The Secretary- General shall thereupon communicate any proposed amendments to the States Parties to the present Covenant [. . .]. In the event that at least one third of the States Parties favors such a conference, the Secretary-General shall convene the conference under the auspices of the United Nations. [. . .]".

been needed for the creation of a new right.[378] According to Tully, the interpretation chosen by the Committee is in contrast to the will of the drafters of the Covenant, who rejected the inclusion of a right to water, and contradicting the prevailing opinion in the literature on the meaning and scope of general comments.

If one engages in a discussion on textual interpretation and the *travaux preparatoires*, as Tully suggests, it means discussing whether the list of sub-rights to the right to an adequate standard of living is "exhaustive" or not. Why did a right to water, unlike the right to food, remain absent in the enumeration of Art. 11 ICESCR? In the words of Matthew Craven, is it an "inclusionary absence", an "exclusionary absence" or an "absence of mind"?[379] We can only speculate about this question, as the *travaux preparatoires* are not entirely clear; the right to water was very briefly considered, but eventually not included.[380] Had a right to water been intensely discussed and then purposefully not been included, one could form an argument in the way Tully does. However, the silence of the drafters is not a strong hint in one or the other direction. When the Covenant was drafted in the early 1960s, the global water crisis was not as urgent as today,[381] and the time to phrase natural goods in human rights terms had not ripened.

Thus, the most convincing option is to abstain from forming an argument from the *travaux preparatoires*: it would not be a reliable argument, either way. This approach is also backed up by the order of interpretative tools established by the Vienna Convention on the Law of Treaties which proscribes that a treaty shall be interpreted in good faith and in accordance with the ordinary meaning to be given to the terms of the treaty in their context and in the light of its object and purpose, and that subsequent agreements, practice and any relevant rules of international law should also be taken into account.[382] Art. 32 of the VCLT clearly states that, *only* if these interpretative tools produce a meaning that is ambiguous, obscure, manifestly absurd or unreasonable, should the *travaux preparatoires* also be used as interpretative aid. That all the primary interpretative tools have failed, or would produce an absurd result, has not been shown by Tully, as also Langford emphasizes.[383]

Tully's second part of this argument—that the establishment of a new right extends the meaning of General Comments beyond what they were intended to be—is to be taken seriously. Some other authors have expressed similar criticism of General Comments. In particular, Michael J. Dennis and David P. Stewart have claimed that the Committee takes "an aggressive view of enforceable rights",[384] and that, in elaborating specific elements under Art. 11 ICESCR, the Committee has

---

[378] Tully (2005), p. 35.

[379] Craven (2006), p. 37.

[380] Riedel (2005), p. 595 (fn. 34); Riedel (2006), p. 24 (fn. 19).

[381] Riedel (2006), p. 19, who concludes also that the travaux préparatoires are not an argument against the assumption of new rights.

[382] VCLT, 23 May 1969, UNTS 1155 (1980), 331, Art. 31(1) and (3).

[383] Langford (2006a), p. 433.

[384] Dennis and Stewart (2004), p. 462.

overridden the decisions of the treaty negotiators and taken an position inconsistent with the view of Member States.[385] For those authors, the right to water is the peak of the Committee's unjustified activism: they call it "virtually without precedent" and a "unilateral alteration in the substantive content of the Covenant".[386] The Committee, according to Dennis and Stewart, had no "authority to rewrite the provisions of the Covenant" in the way it did.[387]

Of course, the acceptance of new rights is ultimately in the hands of the States, not in the hands of an expert committee. It is the main task of the Committee's General Comments to interpret the Covenant in order "to develop a fuller appreciation of the obligations of State Parties under the Covenant".[388] Therefore, the General Comments' main function is "to assist and promote the further implementation by States parties of the Covenant" and to make "the experience gained through examination of States parties' reports available for the benefit of all State parties".[389] However, as was stated before, one must be reluctant to read the drafters' silence on the right to water in the one or the other direction. What remains of Dennis' and Stewart's criticism, is then only the claim that the Committee exceeded its powers. However, the Committee certainly recognizes that it is not in a position to create new rights. The Committee's argument goes more in the direction that related rights clearly accepted under the Covenant cannot be fulfilled without realizing certain individual claims to water. The Committee has interpreted the Covenant, while remaining careful not to exceed its powers. Had the Committee claimed their findings to be legally binding, in a way that the General Comment was the source of a new right itself, Dennis' and Stewart's criticism would be valid. However, the General Comment only suggests that the right is already existent (not to be newly created)[390] and that its source is not the General Comment, but already existing Covenant rights. If this was exceeding the General Comments' power, these powers would be marginalized to a degree that would make their function meaningless. The critique that authors like Dennis and Stewart express is not mirroring a view on General Comment No. 15 or a human right to water, but rather the general criticism towards economic and social rights and their enforceability. In Dennis' and Stewart's own (revealing) words, "the rights and obligations contained in the ICESCR were never intended to be susceptible to judicial or quasi-judicial

---

[385] *Ibid.*

[386] *Ibid.*

[387] *Ibid.*

[388] ECOSOC, *International Covenants on Human Rights*, E/RES/1990/45, adopted on 25 May 1990, par. 10.

[389] UN Commission on Human Rights, *Question of the realization in all countries of the economic, social and cultural rights contained in the Universal Declaration of Human Rights and in the International Covenant on Economic, Social and Cultural Rights, and study of special problems which the developing countries face in their efforts to achieve these human rights*, E/CN.4/RES/2003/18, 22 April 2003, par. 11(a)(ii).

[390] Riedel (2006), p. 19.

determination".[391] What Dennis and Stewart criticize is not the content of the right to water or the approach taken in General Comment No. 15: it is in truth a critique of treating economic and social rights as rights at all.

Lack of an Institution to Oversee the Right?

Tully's third critique of General Comment No. 15 is the omission of an institution, or several institutions, responsible for the oversight of water (such as a specialized UN institution).[392] Tully in fact uses the non-existence of such an institution as (another) argument that the omission of water in the Covenant was intentional.

As argued above, the omission itself is not a fruitful starting point for discussion. However, behind Tully's argument lies a valid question: is it effective to create a right for which the responsible institutions are not clearly defined? The question is then, however: should this absence of a competent institution have stopped the Committee from establishing the right, or should it instead have stated more clearly which institution, in the Committee's view, would be responsible?

Firstly, as also Langford in his rebuttal of Tully's critique rightly points out, other rights under the right of an adequate standard of living have either received institutions responsible for their oversight later than the establishment of the right or have never received a responsible agency.[393] Secondly, in today's institutional set-up, there are many institutions responsible for the oversight and implementation of water rights within the UN system, coordinated under the roof of UN Water.[394] The number of responsible agencies is still growing. The recent introduction of the mandate of the Independent Expert on human rights issues related to safe water and sanitation, in the meantime named Special Rapporteur, shows that the UN is constantly working on giving water a procedural framework. Even if Tully's argument was considered valid in 2002, it is certainly outdated in 2013.

Other Sources for a Right to Water?

Fourthly, Tully argues that there was no need for a new recognition of the right to water, and that, even if one favored such a right, other rights would have been more appropriate sources than the chosen ones.[395] These other rights, such as the rights to housing, food and health, cover all essential water issues, so Tully argues. This was reflected in former General Comments, the mandates of special rapporteurs working in related fields, and through the examples of many national legal orders.[396]

---

391 Dennis and Stewart (2004), p. 462.

392 Tully (2005), p. 35.

393 Langford (2006a), p. 433.

394 UN Water, founded in 2003, coordinates all UN initiatives related to water, see for an overview http://www.unwater.org.

395 Tully (2005), p. 35.

396 *Ibid.*

Tully is partly right here: to anchor the right to water with the right to an adequate standard of living, supported by the right to health, was one, but not the only possible way to approach the topic of water in human right terms. Other rights such as life and human dignity, food and housing, as well as the right to be free from hunger (Art. 11(2) ICESCR) have great relevance for the issue of access to water. However, to argue that a more explicit right to water was not necessary appears almost cynical. The pressing water needs of the people as well as the need of the private sector for clear legal regulation are so multiple and specific that answers to all those questions cannot be found sufficiently through other rights. If several UN special procedures were to deal with the topic, it would take much effort to coordinate these efforts, and none of the mandate holders could deal with the topic in depth, as the workload of their unpaid mandates is already burdensome.[397] A topic as complex and as important as that of water certainly deserves explicit recognition and a serious attempt to describe its normative content, just the way the General Comment did.

States Unwilling to Accept the Right?

Furthermore, Tully argues that State practice does not prove that States are yet willing to accept a human right to water and that their reaction to the General Comment was "lukewarm".[398] In particular, the explicit disagreement of some powerful governments with the General Comment, such as the USA,[399] indicates insufficient consensus on a right to water in international law. Thus, even if it was in the general power and competence of the Committee to release the General Comment, so Tully, the time for establishing this particular human right was not ripe and the Committee's attempt to accelerate the acceptance of the right failed.

This book deals with the status of the right in international customary law in a separate chapter.[400] However, with respect to national legislation, Tully rightly points to those examples of countries where courts deal with essential water issues under the umbrella of the rights to health, housing or food. However, at the same time he leaves other national examples unmentioned, where the right to water has found direct and explicit constitutional or other legal recognition. General Comment No. 15 has surely influenced, if not inspired, the inclusion of a right to water into some national constitutions in the years after its release.

[397] See Sect. 4.1 of this book, in particular remaining problems at 4.1.3.

[398] Tully (2005), p. 35.

[399] "With respect to General Comment No. 15 of the Committee on Economic, Social and Cultural Rights, the United States notes that it does not share the view of the Covenant expressed in that document.", Explanation of Vote, Statement delivered by Joel Daniels, US Delegation to the 61st Commission on Human Rights, 15 April 2005, quoted after Langford (2006a), p. 433.

[400] See Sect. 2.3.3 of this book.

A Right Without Practical Relevance?

Finally, Tully claims that the General Comment does not make any difference in practice and does not put emphasis on the most crucial water needs. He points out that the General Comment focuses too much on "access" to water, and limits itself on States as addressees. Some of the key players crucial to improving the global water situation, the multinational water corporations, remain unaddressed. The General Comment wrongly assumes, so Tully, a State-centric model of international law[401] and is "skewed against governments".[402] The General Comment should have taken into account the real scenario in the water sector of today, namely that of increasing privatization of water supply systems around the globe. In this respect, Tully furthermore critiques that General Comment No. 15 should made clearer how it positions itself in terms of environmental law and water as an environmental resource.[403] According to Tully, the General Comment should also have called upon individuals to stop water wastage.[404]

Tully is right in pointing out the key role of privatization and Foreign Direct Investment in the water sector.[405] Without addressing the question of private sector involvement, a right to water cannot be effectively realized. However, it is understandable that the Committee focused on identifying the obligations of governments, and only highlighted a few obligations of the private sector. It is in part absurd to claim that the General Comment exceeded its authority in terms of content (by establishing a new right, as Tully would call it), before then claiming the Committee should have exceeded its competence even further with respect to its addresses (the private sector rather than States). General Comments are neither addressed directly to the private sector nor towards individuals. They are tailored towards States: to give them guidance and demand of them to stand by the human rights commitments they have themselves agreed upon.

Conclusion: General Comment 15's Role and Significance

While Tully's critique is rather extensive and covers various negative aspects with respect to competence, procedure and content of the General Comment, Tully neglects to name those elements of the General Comment that are praiseworthy. This would have made his contribution a real "critique". The Committee has been very specific in elaborating on the normative content of a human right to water. It has not restricted itself to formulating vague claims or paying political lip service. It made an effort to be consistent with the content and language of former General

---

[401] Tully (2005), p. 35.

[402] *Ibid.*

[403] *Ibid.*

[404] *Ibid.*

[405] This book dedicates a whole sub-chapter to this topic, see Sect. 4.2.2.

Comments, focusing on the main problems and naming those entities responsible, elaborating their responsibilities. In this, the Committee has done very valuable and difficult work. The Committee has relied on the work of other organizations, in particular the recommendations and guidelines developed by the World Health Organization, used its credibility as a body of experts rather than a political institution, and taken into account scientific knowledge to underpin its claims. Altogether, the General Comment, even without legal bindingness, has been a milestone on the way to the establishment of a right to water, or, as chairman Eibe Riedel would phrase it, an important basin in the "cascade model"[406] of international law-making.

However, in some respects the General Comment has, maybe surprisingly, also done a disservice to the right to water. It is less a disservice that has to do with the work of the Committee, but rather with the way it has been perceived ever since. Since 2002, the discussion on the right to water has focused on an adequate standard of living almost exclusively. The Committee mentioned Art. 11 and Art. 12 as two norms of special relevance for the right to water. Even more, it has been neglected that the Committee also mentioned in the General Comment the relation to other highly fundamental rights and legal notions, such as the right to life and human dignity. Already in paragraph 1 of the General Comment, the Committee emphasizes that the human right to water is indispensable for a life in "human dignity". The right as a whole should be understood in conjunction with other rights, foremost amongst them the right to life and human dignity (par. 3). All elements of the right to water must be adequate not only for human health, but also for human dignity and life (par. 11). That it could not elaborate further on them, was a consequence of the Committee's given mandate and function. It was not the implicit suggestion that these rights were less relevant for identifying the legal anchor for a human right to water.

2.3.3 *Protection of the Right to Water through International Custom*

We have so far noted that it takes some effort to construct a human right to water as part of international treaty law. The main shortcomings are that there is no self-standing recognition of such a right that is at the same time legally binding and comprehensive for all human beings. The next logical question is whether a right to water exists as part of international customary law.

As international custom, the right could even have a potentially broader reach: it would include legal effects also for those States that have not signed the ICESCR or

[406] Riedel (1986), p. 164; *id.* (1991), p. 58; more specifically for ESCR-Committee, *General Comment No. 15*, *id.* (2006), p. 19.

have not ratified it,[407] amongst them important States like the USA. Generally speaking, customary law arises if adherence to a norm is reflected in a general and consistent *State practice* and in a corresponding *opinio juris*—the latter being evidence of States' recognition to accept a certain practice as law.[408]

Nothing much apart from the requirement of these two general elements is clear in the concept of international custom. In particular, the classification of border cases (such as "statements" as being either State practice or *opinion iuris*) is highly contentious. This chapter assumes, as explicitly suggested by Anthony A. D'Amato[409] and famously contested by Michael Akehurst,[410] that State practice mainly corresponds with a State's action, whereas *opinio iuris* can mainly be derived from a State's public statements: thus, treaties, declarations and statements fall in the category of evidence for a State's legal conviction (*opinio iuris*) rather than in the category of its action (State practice).[411]

The country studies of this chapter have already indicated that an approach representing a consistent *opinio iuris* and State practice with regard to water is not easily identified. States (do or do not) recognize and give legal effect to a human right to water in very different ways. As for the State practice, in order for it to be relevant of international custom, such a State practice would need to exist independent of the duties from the ICCPR and ICESCR.[412] Additionally, a relevant number of States have so far abstained from signature and ratification of the ICESCR. Thus, even if the right to water was to be considered part of the ICESCR, this would not give it the status of international customary law: the ICESCR itself can currently not be considered part of customary law in all its aspects.[413]

This was safe to say about the customary law status of the right until a few years ago. Four recent events, however, could have changed this clear finding and could have elevated the human right to water above some of the other rights in the ICESCR. These events might be considered as recent manifestations of an arising (or already arisen) *opinio iuris* and/or consistent State practice with respect to water: firstly, the Declaration of Millennium Development Goal (MDG)

---

[407] 160 States are party to the ICESCR. For a list of the current status of signatures and ratification of the ICESCR, see http://treaties.un.org/pages/ViewDetails.aspx?src=TREATY&mtdsg_no=IV-3&chapter=4&lang=en.

[408] Statute of the International Court of Justice, Art. 38(1)(b). See Crawford (2012), pp. 23 ff.; Heintschel von Heinegg (2004), pp. 211 ff.; Shaw (2008), pp. 72 ff.

[409] D'Amato (1971), p. 160; supported for instance by Roberts (2001), p. 758.

[410] Akehurst (1975), pp. 2, 35.

[411] See Roberts (2001), p. 758; D'Amato (1971), p. 89 (albeit arguing that at least treaties should be seen as action).

[412] Rudolf (2007), p. 15.

[413] Even Alston (2005), p. 755, suggesting a more expansive approach to economic and social rights as part of customary law, admits that "it cannot be denied that part of that corpus of rights continues to be the subject of strong resistance on the part of many governments, at least in practice"; also see Rudolf (2007), p. 15; McCaffrey (1992), p. 1.

No. 7 to halve the population living in global thirst by 2015[414]; secondly the UN General Assembly Resolution of July 2010 on "The human right to water and sanitation"[415]; thirdly the HRC Resolution on "Human rights and access to safe drinking water and sanitation" of September 2010[416]; fourthly, the recent re-appointment of the Independent Expert on issues relating of human rights obligations relating to safe water and sanitation, and the renaming of her mandate into "Special Rapporteur on the human right to water and sanitation".[417]

2.3.3.1 Millennium Development Goal Number 7

Philip Alston has argued in his discussion of the MDGs that at least some of the goals have reached the status of international customary law. He suggests to employ two conditions: a right should be considered customary international law, if the norm embodied in one of the development principles is indispensable to a meaningful notion of human dignity—the value inherent in all human rights' protection—and if, secondly, its realization is within the reach of a State (potentially with other States' support).[418] Alston argues that many of the MDGs would meet these conditions, in particular MDGs 1–6; MDG 7 would at least partly be a "strong candidate" for recognition as customary law.[419] Christian Tomuschat has proposed a similar, but also different approach. He proposes to slightly deviate from the common two conditions of customary law and suggests instead that a norm should be recognized as customary law if the norm is necessary to protect human life and physical integrity, because otherwise "the entire idea of a legal order would collapse".[420] In line with this argument, one could make the case for some of the MDGs to be so essential for human life and physical integrity that they should be part of international customary law.[421]

Alston is right insofar that there is no automatism: the fact that economic and social rights are not entirely part of customary law and that the MDGs relate closely to economic and social rights does not exclude MDGs from having the status of customary law. The MDGs are a kind of "minimum selection" of economic and social rights, much narrower in their focus but therefore more widely accepted. This

---

[414] UNGA, *United Nations Millennium Declaration*, A/RES/55/2, 18 September 2000, par. 19.

[415] UNGA, *The human right to water and sanitation*, A/RES/64/292, 3 August 2010.

[416] HRC, *Human rights and access to safe drinking water and sanitation*, A/HRC/RES/15/9, 6 October 2010

[417] See HRC, *The human right to safe drinking water and sanitation*, A/HRC/RES/16/2, 8 April 2011, par. 4.

[418] Alston (2005), p. 755.

[419] *Ibid.*

[420] Tomuschat (2004), p. 35.

[421] However, one should emphasize that Tomuschat himself solely intended civil and political rights to satisfy this requirement, e.g. the prohibition of torture, the protection of personal freedoms or the abolition of slavery.

makes them indeed somehow different from economic and social rights in general.[422] However, as much as it might be a desirable result to expand the reach of customary law onto the MDGs in order to strengthen State accountability for them, it remains unclear why the fact that States have recognized their general ability and their intent to realize a certain development goal by a certain timeframe would be equivalent to their recognition of a legal obligation to do so. The recognition that a norm can, and should, be implemented is different from the recognition that the norm must be implemented, and that legal consequences will be accepted if they are not. It might also be problematic to accept a human right to water through the backdoor of MDG No. 7 as customary law for the future prospects of the MDGs. It might create reluctance on behalf of the States to commit themselves to political goals in the future, if these goals were later on turned into legal obligations. If the MDGs were given the status of customary law, it would be unlikely for many States to agree on follow-up goals to the MDGs in the future.[423] Thus, MDG No. 7 in itself should not be regarded as a suitable proof for a human right to water as part of international custom. It is rather another step on the way of the emergence of such a right in international custom.

2.3.3.2 The UNGA Resolution on "The Human Right to Water and Sanitation"

The 2010 declaration of the UNGA on "The right to water and sanitation"[424] has found a respectable echo in politics and the media.[425] However, as was elaborated above,[426] UNGA resolutions are generally legally non-binding: Art. 10 and 14 of the UN Charter refer to UNGA resolutions as "recommendations", so that their recommendatory nature is without doubt. However, UNGA resolutions can nevertheless be expression of a norm of customary law. This is widely accepted for the case of many provisions of the Universal Declaration of Human Rights (UDHR),[427] although much less accepted for many of the economic and social rights enshrined in it, in particular for the right to an adequate standard of living.[428] Thus, the question is whether the 2010 UNGA Resolution on the right to water and sanitation

[422] Alston (2005), p. 755.

[423] See UNGA, *Keeping the promise: united to achieve the Millennium Development Goals*, A/RES/65/1, 19 October 2010. It is doubtful whether such a consensus would have been reached under the assumption of the MDGs to be customary law.

[424] UNGA, *The human right to water and sanitation*, A/RES/64/292, 3 August 2010.

[425] See for instance, maybe most prominently, Gorbachev (2010), who called already 2 weeks before the resolution upon States to support it. See also Gleick (2010); Barlow (2010).

[426] See Sect. 2.3.1 of this book.

[427] Charlesworth (2008), par. 14 ff.; Hannum (1995–1996); Humphrey (1979), p. 21; McCaffrey (1992), p. 1. See also OHCHR, Digital Record of the UDHR, February 2009, http://www.ohchr.org/EN/NEWSEVENTS/Pages/DigitalrecordoftheUDHR.aspx.

[428] McCaffrey (1992), p. 1.

is one of those UNGA resolutions that provide sufficient indication for a crystallized norm of custom?

To answer this question, one has to scrutinize the drafting process, substance and discussion of the resolution in the UNGA. On the positive side, the high rate of approval of the UNGA resolution is remarkable: the resolution was finally passed with 122 nations in favor, 41 abstentions and no nation voting against it, while 29 nations were not present in the assembly.[429] With 192 members of the UNGA at the time, that makes a three quarters majority of present States, equivalent to a two thirds majority of all 192 States, expressing their positive opinion to recognize water and sanitation as human rights. It is in particular notable that not a single State voted against the resolution—even previously declared sceptics of a human right to water and sanitation, like the USA, Canada or the UK, decided to abstain rather than to vote against the resolution. Thus, by their voting behavior, no State explicitly expressed their denial of a human right to water and sanitation. Additionally, the high number of almost 40 co-sponsors[430] emphasizes that there was a high number of States that emphatically favored the resolution.

With regard to the procedure, the resolution was very much a project pushed through by some States, in particular Bolivia. Many States expressed concerns about the lack of an inclusive drafting process,[431] and complained that their suggestions, and legal convictions, had not sufficiently been taken into account.[432] In particular European States largely regretted that the various suggestions made by the EU had not been taken into account at all.[433] A process to find a common *opinio iuris* on the right to water, as informally undertaken within the framework of the HRC under the name of the "Geneva Process",[434] had not taken place in the preparation of the resolution. Some delegations argued that the resolution even

---

[429] See UNGA, *General Assembly adopts resolution recognizing access to clean water, sanitation as human right*, Press Release GA/10967, 28 July 2010.

[430] See UNGA, *Summary of the 108th plenary meeting of the 64th General Assembly*, A/64/PV.108, 28 July 2010, Statement by the Bolivian representative, pp. 4–6, naming all co-sponsoring countries in its presentation of the "historic resolution".

[431] See UNGA, *Summary of the 108th plenary meeting of the 64th General Assembly*, A/64/PV.108, 28 July 2010, pp. 4 ff. Therein, see statement of the Hungarian representative (p. 7) expressing concerns about the negotiating process in general, the statement of the US representative (pp. 7–8), criticizing that the resolution had not been drafted in a transparent and inclusive manner, and the statements of the New Zealand (p. 11) and Japanese (p. 14) representatives, both critiquing that the resolution was tabled too early, without sufficient time for the delegations to consider it properly.

[432] *Ibid.*, see in particular the statement of the Colombian representative (p. 13), claiming that certain proposals were not considered for inclusion and that the draft remained insufficiently discussed.

[433] *Ibid.*, see for instance statements of the German representative (p. 6), the Spanish representative (p. 6) and the Belgian representative (p. 15).

[434] See for a brief overview of the Geneva Process Furch (2010), pp. 33 ff.

endangered such a deliberative process in the future,[435] and that more language of the ongoing Geneva process would have needed to be included in order to reach consensus.[436] Thus, overall there was no deliberative process to find a common *opinio iuris* to be expressed in the UNGA resolution.

The one sided-support of the resolution, rather than its universal acceptance, becomes also apparent by examining the list of States co-sponsoring the resolution: it is not a cross-section of all UN States. The resolution is largely co-sponsored by developing States, most of them African or South American. No North-American State and no Member State of the EU co-sponsored the resolution.

With regard to the discussion in the UNGA and the voting behavior, one must firstly note the relatively high number of absent votes (29 delegates were absent) and the high number of abstentions (a quarter of all delivered votes). In order to assume a shared *opinio iuris*, a higher number of positive votes would have been needed. The bare fact of no negative votes does not reflect sufficiently the high number of not delivered votes and the even higher number of abstentions. This certainly had to do with the before mentioned Geneva Process that which western States did not want to endanger by a somehow unexpected UNGA resolution. However, silence, be it through abstention or absence, cannot without cautious scrutiny be equated with positive agreement.[437]

Secondly, having a closer look into specific statements, some of the abstaining States even explicitly made clear that their abstention should not be equated with accepting a human right to water and sanitation. Australia explicitly expressed its concerns about the attempt to declare new rights by the means of a UNGA resolution.[438] The UK emphasized that there was no sufficient basis for recognizing water and sanitation as freestanding human rights through treaty or customary law.[439] The USA argued that the resolution would describe a right that was not reflected in existing international law.[440] Canada added that the text of the resolution was premature, and that a non-binding resolution could not try to set out a right without defining its scope; such a right could simply not be declared.[441] Thus, many

---

[435] UNGA, *Summary of the 108th plenary meeting of the 64th General Assembly*, A/64/PV.108, 28 July 2010, statement of the US representative (pp. 7–8) and statement of the Turkish representative (p. 8).

[436] *Ibid.*, see for instance statement of the UK representative (pp. 12–13) and statement of the Brazilian representative (p. 5).

[437] See generally D'Amato (1969); MacGibbon (1957). As an example of acquiescence, see ICJ, *Temple of Preah Vihear* (Cambodia v. Thailand), Judgment of 15 June 1962, ICJ Reports (1962), 6, 23 and 31.

[438] UNGA, *Summary of the 108th plenary meeting of the 64th General Assembly*, A/64/PV.108, 28 July 2010, Statement of the Australian representative, p. 11.

[439] *Ibid.*, Statement of the UK representative, p. 12.

[440] *Ibid.*, Statement of the US representative, pp. 7–8.

[441] *Ibid.*, Statement of the Canadian representative, p. 17.

of the State abstentions are, in fact, quasi-negative votes. It almost appears that several States abstained instead of rejecting the resolution only due to the severity of the global water crisis and out of respect to its victims, as very figuratively and explicitly pointed out by the Bolivian representative in his initial statement before the vote.[442] Even some of the States voting in favor of the resolution, such as Guatemala, Egypt and Liechtenstein, expressed their understanding that the resolution did not create any new right or any legal obligation on the international level.[443] Colombia noted that its vote in favor of the resolution only implied a general acceptance of a right to drinking water, not to sanitation.[444]

Thirdly, even the content of the resolution itself does not easily allow for the assumption of a recognition of the right as customary law. The resolution consists of only three operative paragraphs: par. 1 "recognizes" the right to safe and clean drinking water and sanitation as a human right essential for life and all human rights; par. 2 calls upon States and IGOs to invest in international assistance and cooperation for developing countries to accelerate their efforts related to drinking water and sanitation; par. 3 welcomes the work done by the independent expert and encourages her to elaborate the challenges to water and sanitation, and the realization of the MDGs. The restricted language of these three paragraphs emphasizes that no clear content of the right to water and sanitation was outlined. No mention of specific elements of General Comment No. 15 was made (apart from its general mention at the very beginning of the resolution, albeit together with various other documents). No mention is made of the right's legal origins (e.g. the right to an adequate standard of living). Without applying too high standards for the content for a new right, one must nevertheless seriously question what a shared *opinio iuris* is worth, if legal origin, scope and limits of such a right remain undefined.

Interesting in this regard is a last-minute change of the language of the resolution: In operative paragraph 1, the centerpiece of the resolution,[445] the word "declares" was replaced by the word "recognizes". This suggests that the drafters acknowledged that a UNGA resolution could not "declare" a new right (due to a UNGA resolution's non-binding nature), but that it could at most be seen as expression of the "recognition" of an already existing right. However, given all the mentioned hesitant and critical interventions in the course of the discussion, it appears unconvincing to derive either an unanimous *opinio iuris* or a consistent State practice of all UN Member States from the resolution alone.

Thus, the UNGA resolution on "The human right to water and sanitation" of July 2010 has by itself created a right to water and sanitation as part of international customary law.

---

[442] *Ibid.*, Statement of the Bolivian representative, pp. 4–6.

[443] *Ibid.*, Statement of the Guatemalan representative, p. 5; Statement of the Egyptian representative, p. 5; Statement of the Liechtenstein representative, p. 8.

[444] *Ibid.*, Statement of the Colombian representative, pp. 6–7.

[445] The Permanent Representative of Bolivia calls it "the heart of the resolution", see *ibid.*, Statement of the Bolivian representative, pp. 4–6.

2.3.3.3 The HRC Resolution "Human Rights and Access to Safe Drinking Water and Sanitation"

Like the UNGA resolution, the HRC resolution does not have the legal effect of an international treaty and could as such not create a legally binding human right.[446] However, there is a remarkable trend visible from the UNGA resolution in July to the HRC resolution in October.

The fact that the resolution was adopted without a vote is a great success for and an appreciation of the work of the Special Rapporteur. Nations that were critical towards previous resolutions, such as the USA or the United Kingdom, were generally positive towards the HRC resolution in their statements.[447] Individual States might have changed their attitude towards the right to water and their respective practice during the Geneva Process. Remarkable in this respect are in particular the statements of the US and British representatives, who suggest that their countries were generally willing to give up their previous opposition to the right to water.[448] Only a few months after the UNGA resolution, in which many States had still abstained from voting, States now expressed a unanimously positive attitude towards the right. Secondly, also in terms of content, the resolution was much more concrete than its predecessor resolution of the UNGA: it suggested the right to be derived from the right to an adequate standard of living, while also emphasizing important links to other human rights.

The pitfalls of the resolution, however, must not be forgotten. A norm of international custom, thus a shared *opinio iuris* amongst all States and a consistent State practice, cannot be evidenced by a resolution of a body with only 47 members alone (e.g. not all States are represented within that body). Forty seven States cannot speak for all (at the time) 192 States; not all of them can be considered to be legally bound. Even more, if one allowed a single resolution to create international custom, this would give the resolution the effect of "instant custom",[449] which the ICJ has explicitly rejected.[450] Custom is more a growing process than a single event.

---

[446] See Sect. 2.3.1 of this book.

[447] See the statements of the representatives in the archived webcast of the HRC, http://www.un.org/webcast/unhrc/index.asp.

[448] UNGA, *Summary of the 108th plenary meeting of the 64th General Assembly*, A/64/PV.108, 28 July 2010, statement of the US representative (pp. 7–8), who expresses happiness that his country can join in the unanimous resolution. However, the UK representative expressively disagrees with respect to that part of the resolution which relates to sanitation (which, however, lies outside the scope of this study).

[449] Cheng (1965); D'Amato (1971). But see Akehurst (1975), pp. 12 ff., 31 ff.; Shaw (2008), pp. 78–79; Heintschel von Heinegg (2004), pp. 214 ff.

[450] ICJ, *North Sea Continental Shelf* (Federal Republic of Germany v. Denmark; Federal Republic of Germany v. Netherlands), Judgment of 20 February 1969, ICJ Reports (1969), 3, 43; ICJ, *Case Concerning the Military and Paramilitary Activities In and Against Nicaragua* (Nicaragua v. USA), Judgment of 27 June 1986, ICJ Reports (1986), 14, 97.

Is the resolution thus worthless? Can HRC resolutions never create new custom? Things are not black and white. There is a clear trend visible from the UNGA resolution of July 2010 and the HRC resolution of October 2010. Some of those States being represented in both bodies, and being present at both votes, changed their voting behavior. Thus, the HRC resolution marked a new clear trend: the strict division between those States in favor and those opposed to the right has begun to get blurred. It thus appears preferable to see the resolution in a light that Eibe Riedel already suggested for General Comment No. 15 some 10 years ago, namely to see it as an important next step along the "cascade model" of international law-making.[451] The HRC resolution was certainly a milestone, or an important cascade, in establishing such a norm of international custom.

2.3.3.4 Renewal of the Mandate of the Independent Expert as Special Rapporteur

If the main critique of the HRC resolution is that only 47 Members could not establish a new norm of custom for all States, the reaction that the resolution received by the broader international community is of crucial importance. Custom cannot only be created by active (voting) behavior, but by passively accepting and not protesting against a resolution,[452] in particular if they are of global significance as UNGA and HRC resolutions certainly are.

There were no mayor objections in the aftermath of the HRC resolution traceable. States welcomed the resolutions, not at all protesting against them. To the contrary, States even took another affirmative step towards the recognition of a human right to water: when the mandate of the "Independent Expert on the issue of human rights obligations relating to access to safe drinking water and sanitation" was up for renewal after 3 years in spring 2011, States decided not only to renew the mandate (which was certainly to be expected and in itself would not be a hint for increased recognition of the right), but also to give the mandate a new name. It was renamed "Special Rapporteur on the human right to water and sanitation". There might not be a legally significant difference in the title of "Independent Expert" and "Special Rapporteur"; however, it is clear that the subject of the mandate was now termed directly in human right terms (in comparison to the complicated original title of 2008 which willfully avoided the term "human right to water and sanitation".) Half a year after the HRC Resolution and almost a year after the UNGA Resolution, opposition to the term "human right to water and sanitation" had been given up.

Also in terms of phrasing the Special Rapporteur's tasks, a feasible change is apparent. While part of the first mandate had been to elaborate the content of human right obligations relating to water ("clarification of the content of human rights obligations, including non-discrimination obligations, in relation to access to safe

[451] Riedel (1986), p. 164.

[452] See D'Amato (1969); MacGibbon (1957); Shaw (2008), pp. 89 ff.

drinking water and sanitation"),[453] the Special Rapporteur's renewed mandate is now on questions of practical implementation of the right ("promote the full realization of the human right to safe drinking water and sanitation by, inter alia, continuing to give particular emphasis to practical solutions with regard to its implementation"[454] and "to work on identifying challenges and obstacles to the full realization of the human right to safe drinking water and sanitation, as well as protection gaps thereto, and to continue to identify good practices"[455]). The language of the HRC has thus clearly moved within three years from "human rights obligations relating to access to water and sanitation" to the "human right to water and sanitation". That the right exists as part of a shared *opinio iuris*, seems accepted for the HRC and its members by now.

2.3.3.5 Assessment of the Current Situation

These four events (increasing in legal significance according to the order listed above) give a more or less solid basis to the claim that a shared *opinion iuris* has finally arisen. The question which remains is that of State practice. It is necessary that the Special Rapporteur identifies in the years to come to what extent State practice is consistent or moving in that direction. Some authors have so far concluded that it is in particular State practice which currently still bars the right from being part of international custom.[456] While such an approach is certainly not incorrect, one might question whether a more flexible, if not modern, view of the interplay of *opinio iuris* and State practice would come to a different conclusion here. Thus the question is which element, if any, is of greater importance in the determination of a rule of international custom.

While most authors still assume both elements, *opinio iuris* and State practice as indispensable, many of them suggest a more flexible approach. As Anthea Roberts famously categorizes it, one can distinguish two main streams in this respect: "traditional" and "modern custom".[457]

In this terminology, "traditional custom" heavily relies on State practice—a practice which must be general and consistent and carried out from a sense of

---

[453] HRC, *Human rights and access to safe drinking water and sanitation*, A/HRC/RES/7/22, adopted on 28 March 2008, par. 2(b).

[454] HRC, *The human right to safe drinking water and sanitation*, A/HRC/RES/16/2, 8 April 2011, par. 5(a).

[455] *Ibid.*, par. 5(c).

[456] Rudolf (2007), p. 15; also see Kirschner (2011), p. 445.

[457] Roberts (2001), p. 758 (and all through her essay). What Roberts calls "modern" custom, is very similar to what before her was termed "contemporary" custom (Stein (1988), pp. 12–13) and "new" custom (Bradley and Goldsmith (1997), p. 838).

having a legal obligation to do so (in order to distinguish it from pure habit). The ICJ is often assumed to be a supporter of this approach in its *North Sea Continental Shelf*-case,[458] but also in various other judgments.[459] Thus, by the means of an inductive process a norm of custom is inferred from an observed State practice,[460] whereas *opinio iuris* is rather a secondary element.

"Modern custom" relies on *opinio iuris* as the decisive element for the formation of international custom: the deductive process starts with examining general statements of rules rather than with identifying a certain practice.[461] To what extent these statements evolve into rules of custom depends on the number of supportive States, the exact wording (obligatory or recommendatory character) and on the degree to which the statement is followed by a subsequent State practice.[462] A prominent example from the ICJ case law is the merits decision in the *Nicaragua*-case[463] where the court mainly relied on a UNGA resolution to construe rules of custom relating to the prohibition of the use of force and non-intervention.[464]

These two approaches to international custom have both been extensively criticized. The general criticism for both approaches is that each of them relies too heavily on one element of international custom while neglecting the other. The diverging paths of the two different approaches have, according to some authors, undermined the concept of custom as a reliable and legitimate source of international law as a whole,[465] and have led to an "identity crisis" of international custom.[466] The approaches have also been critiqued individually. "Traditional custom" has been termed a "dinosaur" approach which anachronistically fails to picture the rapidly developing international law.[467] Clear rules of what uniformity, frequency, duration and continuity of the needed State practice must involve do not exist.[468] Even more, to observe State practice of all 193 UN Member States, or analysing a sample that does not have a selection bias and would thus be

[458] ICJ, *North Sea Continental Shelf* (Federal Republic of Germany v. Denmark; Federal Republic of Germany v. Netherlands), Judgment of 20 February 1969, ICJ Reports (1969), 3, 44 ff.

[459] ICJ, *Right of Passage over Indian Territory* (Portugal v. India), Judgment of 12 April 1960, ICJ Reports (1960), 6, 42–43. See also PCIJ, *S.S. "Lotus"* (France v. Turkey), Judgment of 7 September 1927, PCIJ Ser. A No. 10 (1927), 28.

[460] See in particular Schwarzenberger (1947), pp. 566–570; *id.* (1965), p. 33; for more contemporary assessments, see Chodosh (1991), p. 102; Schachter (1991), pp. 35–36.

[461] See on this Simma and Alston (1988–1989); Cheng (1965).

[462] Charney (1993), pp. 544–545; also see Akehurst (1975), pp. 6–7; Roberts (2001), p. 758.

[463] ICJ, *Case Concerning the Military and Paramilitary Activities In and Against Nicaragua* (Nicaragua v. USA), Judgment of 27 June 1986, ICJ Reports (1986), 14.

[464] See for instance UNGA, *Declaration on Principles of International Law concerning Friendly Relations and Co-operation among States in accordance with the Charter to the United Nations*, Res. 2625 (XXV), A/8028, 24 October 1970.

[465] Kelly (2000), p. 451.

[466] Simma and Alston (1988/1989), pp. 88, 96.

[467] Fidler (1996), pp. 216–231; see similar, Charney (1993), p. 543.

[468] Byers (1999), pp. 156–162; D'Amato (1971), pp. 56–66.

representative, is difficult if not impossible.[469] If the analysis focuses on "actions", this neglects States without the ability to act, without sufficient factual knowledge or without the awareness that action or inaction, if it amounts to acquiescence, can also have legally binding effects.[470] This way, some States are bound by (customary) rules in whose creation, knowingly or unknowingly, they were not actively involved.[471]

It might still take time until one can conclude that a sufficient State practice is established, if one applies strict standards to the issue of State practice. However, arguably one should not expect too much in terms uniformity of this practice. Firstly, the water needs are very diverse in the different countries and cultures of the world. Secondly, and in line with the more modern approaches to international custom mentioned above, the strong *opinio iuris*, strengthening feasibly year by year, can to a certain extent make up for the somewhat less clearly established State practice. Ultimately, it cannot lie within the reach of a legal analysis to empirically assess the existing State practice. It is, *inter alia*, for the Special Rapporteur to show that such practice in the different countries is existent and consistent enough, or not, in order to call the right part of customary international law.

Thus, to judge on the question of a human right to water as international custom, the decisive question is how much flexibility one allows in the interplay of State practice and *opinio iuris*. The latter is certainly much more clearly traccable in international law. If one considers, along more traditional approaches to customary law, that existing State practice is currently not sufficiently uniform, one must at least conclude that a strong, and ever strengthening, *opinio iuris* in favor of the human right is currently paving the way for State practice to develop along these lines in the future.

2.3.4 Protection of a Human Right to Water through General Principles of Law

An often forgotten element in the discussion of possible sources for a human right to water in international law is that one of general principles of law, as recognized by civilized nations, according to Art. 38(1)(c) of the ICJ Statute. National constitutions can, if read together with soft law such as international declarations, be an expression of a general principle of international law.[472] General principles of law are therefore expressions of universal values, reflected both in the international

---

469 Charney (1993), p. 537; Chodosh (1991), p. 102; Fidler (1996), pp. 203, 217; Weisburd (1988), p. 6.

470 Byers (1999), p. 37; Kelly (2000), p. 522; Roberts (2001), p. 767.

471 Kelly (2000), p. 522.

472 Kadelbach and Kleinlein (2006), p. 235; Rudolf (2007), p. 15.

and national legal orders.[473] However, in order to assume a value of that universal status, a right to water would need to be widely reflected in national primary and secondary law-making. Given the explicit recognitions of a right to water in only a few national constitutions and ordinary law—some of which were examined earlier in this chapter—such a universal value hinting to a general principle is not yet sufficiently reflected in national law.[474]

2.4 Conclusion

We have seen in this chapter that the protection and the promotion of the right to water has taken very different shape and form in the legal orders across the globe. The differences have various dimensions and fall within different categories: explicit recognition versus derivative recognition; recognition as a subjective right versus recognition as an expression of objective general principles; recognition as a negative right versus recognition as both a positive and negative right; recognition as a socio-economic right versus recognition as a partly civil and political right.

These different approaches are due to political and legal constraints that legislators and judges have to face in responding to the needs and particularities of the legal systems in which they operate. The German courts went back to the constitutional principle and combined it with human dignity, as the latter one is the most important legal concept in the German constitution. The Indian courts referred back to the right to life as a legal anchor for the right to water as only civil and political rights are enforceable under the Indian constitution. The ESCR- Committee, on the other hand, created a purely socio-economic right to water as this was the only right possible within their interpretative powers. Other countries, such as Belgium or France, recognized a right to water directly in their provincial or federal laws. South Africa did so even on the constitutional level, as this was much in line with the spirit of the new constitution the country gave itself in the 1990s.

In international law, many legal documents mention a right to water, but none of them establishes such a right in a legally binding (going beyond political declarations), comprehensive (valid for all humans) and self-standing (thus independent) manner. States have simply been unable to reach consensus on these issues yet. However, recent developments, such as the UNGA and the HRC resolutions as well as the renewal of the Independent Expert's mandate are encouraging signals. They are important steps in the "cascade model" of international law-making.

In the absence of a successful political initiative, it was the judges and quasi-judges who have taken action. Although no consistent line is visible in the different judgments of the different courts, linking the right to water sometimes to civil and political rights and sometimes to economic, social and cultural rights, the courts have at least set the ball rolling for the acceptance of the right in general. In order to act on this suggestion and at the same time to resolve the disorder caused by the

---

[473] Kadelbach and Kleinlein (2006), p. 235.

[474] Coming to the same conclusion: Rudolf (2007), p. 15.

right's ill-defined status,[475] the ESCR-Committee released *General Comment No. 15* on the "Right to Water" in 2002. It therein declared that the right to water was already implied in the ICESCR's right to an adequate standard of living and the right to the highest attainable standard of health. However, General Comments are authoritative interpretative tools, not legally binding in themselves. Nevertheless, General Comment No. 15 represented major progress towards a right to water as a human right. In this regard, one cannot pay enough tribute to the Committee's courage and initiative in this regard. However, the General Comment focuses on the socio-economic dimension of the right, only mentioning peripherally the right to life and the notion of human dignity. The reason for this is obvious: The ESCR-Committee only supervises the interpretation of the ICESCR and has no control over the interpretation of civil and political rights, such as the rights of the ICCPR. Therefore, the Committee did everything possible within its mandate and power. If one accepts a derivative right as a right of its own, one can conclude that the right exists in international treaty law, but one would have to attribute, as Amanda Cahill suggested, a "unique" status to it.

The right is also a "strong candidate" to be considered part of international customary law. If one accepts a certain flexibility between the two elements *opinio iuris* and State practice, as modern approaches to the notion of custom suggest, one can conclude that the right, despite State practice remaining much less evident than *opinio iuris*, has become part of international custom, in particular since the 2010 UNGA and HRC resolutions and the renewal of the Special Rapporteur's mandate, during which States clearly assumed the right to water to have become part of international law.

It is one of the tasks of international human rights law to improve the legal status of the individual, but also reflecting national experiences. When discussing the nature and shape of a human right to water, one should therefore look for an approach that reflects the experiences on different legal levels, both national and international, and which establishes a comprehensive and balanced conception of this right. This is the main task and focus of Chap. 3 of this book.

References

German Federal Foreign Office (AA) (2010) Germany in the United Nations. https://www.auswaertiges-amt.de/cae/servlet/contentblob/382774/publicationFile/4262/DeutschlandindenVN-englisch.pdf

Akehurst M (1975) Custom as a source of international law. Br Yearb Int Law 47:1–53

Alston P (1984) Conjuring up new human rights: a proposal for quality control. Am J Int Law 78 (3):607–635

Alston P (2005) Ships passing in the night: the current state of the human rights and development debate seen through the lens of the millennium development goals. Hum Rights Q 27 (3):755–829

[475] Cahill (2005), p. 389.

Ambrosius G (2000) Service Public, Leistungen der Daseinsvorsorge oder Universaldienste? Zur historischen Dimension eines zukünftigen Elements europäischer Gesellschaftspolitik. In: Cox H (ed) Daseinsvorsorge und öffentliche Dienstleistungen in der Europäischen Union, 1st edn. Nomos, Baden-Baden, pp 15–43
Aquawal (2012) Walloon Drinking Water and Waste-Water Treatment Statistics: 2012 Report. http://www.aquawal.be/servlet/Repository/walloon-drinking-water-and-waste-water-treatment-statistics-2012-report.pdf?ID=4304&saveFile=true
Armeni C (2008) The right to water in Belgium. International Environmental Law Research Center, Briefing Paper 2008-02
Barlow M (2007) Blue covenant: the global water crisis and the coming battle for the right to water. New Press, New York
Barlow M (2010) Access to clean water is most violated human right. http://www.theguardian.com/commentisfree/cif-green/2010/jul/21/access-clean-water-human-right
Baur F (1982) Die "Naßauskiesung" – oder wohin treibt der Eigentumschutz? Neue Jurist Wochenzeitschr 35(32):1734–1736
Bernsdorff N (1997) Positivierung des Umweltschutzes im Grundgesetz (Art. 20a GG). Natur Recht 19(7):328–334
Bluemel EB (2004) The implications of formulating a human right to water. Ecol Law Q 31 (4):957–1006
Federal Ministry of the Environment, Nature Conservation and Nuclear Safety (BMU) (2011) Wasser – Wohlstand – Wandel: Gewässerschutz sichert unsere Grundlagen für Leben, Vielfalt und Nachhaltigkeit. http://www.bmu.de/fileadmin/bmu-import/files/pdfs/allgemein/application/pdf/broschuere_wasser_wohlstand_wandel_bf.pdf
Boury P (2008) Tarification de l'eau en Belgique – Un accès équitable à l'eau. Paper presented at Journée Scientifique et technique ENGEES-CEMAGREF, named Les politiques d'accès à l'eau des public défavorisés mises en ouvre ici et ailleurs: regards croisés. http://www.oieau.fr/IMG/pdf/Boury.pdf
Bowett DW (1957) Estoppel before international tribunals and its relation to acquiescence. Br Yearb Int Law 33:176–202
Bradley CA, Goldsmith JL (1997) Customary international law as federal common law: a critique of the modern position. Harv Law Rev 110(4):815–876
Brinkmeier S (2002) Kontrahierungszwang in der Wasserwirtschaft. Lit Publishing, Münster
Bullinger M (2003) Französischer service public und deutsche Daseinsvorsorge. Juristenzeitung 58(12):597–604
Busche J (1999) Privatautonomie und Kontrahierungszwang. Mohr Siebeck, Tübingen
Busse von Colbe W (2011) Entregulierung für die deutsche Wasserwirtschaft. In: Joost D (ed) Festschrift für Franz Jürgen Säcker zum 70. Geburtstag. C H Beck, Munich, pp 575–587
Byers M (1999) Custom, power, and the power of rules: international relations and customary international law. Cambridge University Press, Cambridge
Cahill A (2005) The human right to water – a right of unique status: the legal status and normative content of the right to water. Int J Hum Rights 9(3):389–410
Castle S (2011) 18 months after vote, Belgium has government. http://www.nytimes.com/2011/12/02/world/europe/belgium-forms-government-with-elio-di-rupo-as-premier.html?_r=0
Charlesworth H (2008) Universal declaration of human rights (1948). In: Wolfrum R (ed) Max Planck encyclopedia of public international law. Oxford University Press, Oxford (online edition)
Charney JI (1993) Universal international law. Am J Int Law 87(4):529–551
Cheng B (1965) United Nations resolutions on outer space: "instant" international customary law? Indian J Int Law 5:23–48
Chodosh H (1991) Neither treaty nor custom: the emergence of declarative international law. Tex Int Law J 26(1):87–124
Center on Housing Rights and Evictions (COHRE) (2008) Legal resources for the right to water and sanitation: international and national standards, 2nd edn. http://www.worldwatercouncil.org/fileadmin/wwc/Programs/Right_to_Water/Pdf_doct/RWP-Legal_Res_1st_Draft_web.pdf
Cornut P et al (2006) Public, "club" and individual management of natural resources: the case of domestic rainwater tanks in Belgium. Die Erde Contrib Hum Geogr 137(4):273–292

Craig P, de Búrca G (2011) EU law: text, cases, and materials, 5th edn. Oxford University Press, Oxford
Cranston M (1967) Human rights – real and supposed. In: Raphael DD (ed) Political theory and the rights of men. Indiana University Press, Bloomington, pp 43–53
Cranston M (1973) What are human rights? Bodley Head, New York
Craven M (2006) Some thoughts on the emergent right to water. In: Riedel E, Rothen P (eds) The human right to water. Berliner Wissenschafts-Verlag, Berlin, pp 37–48
Crawford J (2012) Brownlie's principles of public international law, 8th edn. Oxford University Press, Oxford
Czychowski M, Reinhardt M (2010) Wasserhaushaltsgesetz: WHG Kommentar. C H Beck, Munich
D'Amato AA (1969) Consent, estoppel, and reasonableness: three challenges to Universal International Law. Va J Int Law 10(1):1–31
D'Amato AA (1971) The concept of custom in international law. Cornell University Press, Ithaca
Dennis MJ, Stewart DP (2004) Justiciability of economic, social and cultural rights: should there be an international complaints mechanism to adjudicate the rights to food, water, housing and health? Am J Int Law 98(3):462–515
Department of Water Affairs (2010) Annual Report 2009/2010 (Part 1). http://www.dwaf.gov.za/documents/AnnualReports/2010AnnualReport.pdf
Department of Water Affairs (2012) Annual Report 2011/2012. http://www.pmg.org.za/report/20121010-department-water-affairs-their-annual-report-and-audited-financial-st
Department of Water Affairs and Forestry (2008) Annual Report 2007/2008. http://www.dwaf.gov.za/documents/AnnualReports/ANNUALREPORT2007-2008.pdf
Dreyer J, Bartl U (2010) Preishöhenkontrolle bei Waffenlieferungen – Wasserpreise Wetzlar. Neue Jurist Wochenschr 63(35):2553–2554
Dupuy PM (2006) Le droit à l'eau, un droit international? La mise en oeuvre du droit à l'eau, Institut International de droit d'expression et d' Inspiration Francaises, Actes du XXIXe Congrès ordinaire de l'IDEF, 2006, 29 September – 1st October 2005; at the same European University Institute Working Paper Law 2006/06, Florence
European Committee of Social Rights (ECSR) (2006) Digest of the Case Law. http://www.sm.ee/fileadmin/meedia/Dokumendid/Sotsiaalvaldkond/sotsiaalharta/Case_law_Digest_en_1__1_.pdf
European Committee of Social Rights (ECSR) (2012) Activity Report 2011. http://www.coe.int/t/dghl/monitoring/socialcharter/Presentation/ActivityReport2011_en.pdf
Emmerich-Fritsche A (2007) Privatisierung der Wasserversorgung in Bayern und kommunale Aufgabenverantwortung. Bayrische Verwaltungsblätter 138(1):1–8
Faßbender K (2001) Gemeinschaftsrechtliche Anforderungen an die normative Umsetzung der neuen EG-Wasserrahmenrichtlinie. Neue Zeitschrift Verwaltungsrecht 20(3):241–249
Fidler D (1996) Challenging the classical concept of custom: perspectives on the future of customary international law. Ger Yearb Int Law 39:198–248
Filmer-Wilson E (2005) The human rights-based approach to development: the right to water. Neth Q Hum Rights 23(2):213–241
Forsthoff E (1938) Die Verwaltung als Leistungsträger. Kohlhammer, Stuttgart
Forsthoff E (1959) Rechtsfragen der leistenden Verwaltung. Kohlhammer, Stuttgart
Francis R (2005) Water justice in South Africa: natural resources policy at the intersection of human rights, economics, and political power. Georget Int Environ Law Rev 18(1):149–196
Furch L (2010) Menschenrecht auf Wasser- und Sanitärversorgung: UN-Resolutionen als Schlüssel zum Paradies? Eur J Transnatl Stud 2(2):26–49
Gassner E (2011) Rechtskonkretisierung zum Schutz der natürlichen Lebensgrundlagen nach Art. 20a GG. Natur Recht 33(5):320–323
Geiger R (2010) Commentary on article 11 EUT. In: Geiger R, Khan DE, Kotzur M (eds) EUV/AEUV Kommentar, 5th edn. C H Beck, Munich
Gleick PH (2010) The human right to water (and sanitation). http://www.huffingtonpost.com/peter-h-gleick/the-human-right-to-water_b_671175.html
Goeters H (2008) Das institutionelle Gleichgewicht – seine Funktion und Ausgestaltung im Europäischen Gemeinschaftsrecht. Duncker & Humblot, Berlin

Goldsmith QC (2001) A charter of rights, freedoms and principles. Common Mark Law Rev 38 (5):1201–1216

Gorbachev M (2010) The right to water. http://www.nytimes.com/2010/07/17/opinion/17iht-edgorbachev.html?_r=0

Hall D, Lobina E (2012) Water companies and trends in Europe 2012. Public Services International Research Unit. http://www.epsu.org/IMG/pdf/2012_Water_companies-EWCS.pdf

Hannum H (1995–1996) The status of the universal declaration of human rights in national and international law. Ga J Int Comp Law 25(1–2):287–398

Heintschel von Heinegg W (2004) Kapitel 4: Die weiteren Quellen des Völkerrechts. In: Ipsen K (ed) Völkerrecht, 5th edn. C H Beck, Munich, pp 210–256

Henneke HG (2011) Kommentierung von Art. 28. In: Schmidt-Bleibtreu B, Hofmann H, Hopfauf A (eds) Kommentar zum Grundgesetz, 12th edn. Heymann, Cologne

Herdegen M (2012) Commentary on Art. 1 Abs. 1. In: Maunz T, Dürig G (eds) Grundgesetz: Kommentar. C H Beck, Munich

Herzog R, Grzeszick B (2012) Commentary on Art. 20. In: Maunz T, Dürig G (eds) Grundgesetz: Kommentar. C H Beck, Munich

Human C (2006) The human right to water in Africa: the South African example. In: Riedel E, Rothen P (eds) The human right to water. Berliner Wissenschafts-Verlag, Berlin, pp 83–93

Humphrey J (1979) The universal declaration of human rights: its history, impact and juridical character. In: Ramcharan BG (ed) Human rights: thirty years after the universal declaration. Springer/Kluwer, Dordrecht, pp 21–37

Jaag T (2012) Demokratische Legitimation der EU-Außenpolitik nach Lissabon. Europarecht 47(3):309–322

Jarass HD, Pieroth B (2012) Grundgesetz für die Bundesrepublik Deutschland GG, 12th edn. C H Beck, Munich

Kadelbach S, Kleinlein T (2006) Überstaatliches Verfassungsrecht: zur Konstitutionalisierung im Völkerrecht. Arch Völkerrechts 44(3):235–266

Kämmerer JA (2002) Daseinsvorsorge als Gemeinschaftsziel oder: Europas "soziales Gewissen". Neue Zeitschrift Verwaltungsrecht 21(9):1041–1046

Kämmerer JA (2004) Strategien zur Daseinsvorsorge: Dienste im allgemeinen Interesse nach der "Altmark"-Entscheidung des EuGH. Neue Zeitschrift Verwaltungsrecht 23(1):28–34

Kelly JP (2000) The Twilight of customary international law. Va J Int Law 40(2):449–544

Kersten J (2005) Die Entwicklung des Konzepts der Daseinsvorsorge im Werk von Ernst Forsthoff. Der Staat 44(4):543–569

Kirschner AJ (2011) The human right to water and sanitation. Max Planck Yearb UN Law 15:445–487

Knauber R (1984) Die jüngere Entschädigungsrechtsprechung des BGH nach dem Naßauskiesungsbeschluß des BVerfG. Neue Zeitschrift Verwaltungsrecht 3(12):753–759

Knopp GM (2010) Das neue Wasserhaushaltsrecht. C H Beck, Munich

Köck W (2012) Wasserwirtschaft und Gewässerschutz in Deutschland, Rechtsrahmen, Institutionen, Organisation. Zeitschrift Umweltrecht 23(3):140–150

Kok A (2005) Privatisation and the right to access to water. In: De Feyter K, Isa FG (eds) Privatisation and human rights in the age of globalisation. Intersentia, Antwerp, pp 259–287

Langford M (2006a) Ambition that overleaps itself? A response to Stephen Tully's critique of the general comment on the right to water. Neth Q Hum Rights 24(3):433–459

Langford M (2006b) Expectation of plenty: response to Stephen Tully. Neth Q Hum Rights 24 (3):473–479

Laskowski SR (2010) Das Menschenrecht auf Wasser: Die rechtlichen Vorgaben zur Sicherung der Grundversorgung mit Wasser und Sanitärleistungen im Rahmen einer ökologisch-nachhaltigen Wasserwirtschaftsordnung. Mohr Siebeck, Tübingen

Le Nestour C, Zinow BM (1994) Rechtsfragen des "Service Public". Recht Energiewirtschaft 55(4):129–133 and 55(5):170–174

Leckie S, Gallagher A (2006) Economic, social, and cultural rights: a legal resource guide. University of Pennsylvania Press, Philadelphia

Liebenberg S (1998) The national water bill – breathing life into the right to water. Econ Soc Rights Rev 1(1):3–6

MacGibbon IC (1957) Customary international law and acquiescence. Br Yearb Int Law 33:115–145
MacGibbon IC (1958) Estoppel in international law. Int Comp Law Q 7(3):468–513
McCaffrey SC (1992) A human right to water: domestic and international implications. Georget Int Environ Law Rev 5(1):1–24
McCaffrey SC (2005) The human right to water. In: Bernasconi-Osterwalder N, Brown Weiss E, de Boisson Chazournes L (eds) Fresh water and international economic law. Oxford University Press, Oxford, pp 93–115
McDonald DA (2002) The theory and practice of cost recovery in South Africa. In: McDonald DA, Pape J (eds) Cost recovery and the crisis of service delivery in South Africa. HSRC Publishers/ZED Books, Cape Town/London, pp 17–37
Mehta L (2005) Citizenship and the right to water. Lessons from South Africa's free basic water policy. In: Kabeer N (ed) Inclusive citizenship: meanings and expressions. Zed Books, London/New York, pp 235–250
Mestmäcker EJ (1998) Daseinsvorsorge und Universaldienst im europäischen Kontext: Ein Beitrag zur Funktion von Art. 90 Abs. 2 EGV. In: Ruland F, von Baron Maydell B, Papier HJ (eds) Verfassung, Theorie und Praxis des Sozialstaats. Festschrift für Hans F. Zacher zum 70. Geburtstag. C F Müller, Heidelberg, pp 635–651
Miquel MG (2003) Report to the Senate No. 215 (2002–2003): La qualité de l'eau et l'assainissement en France. http://www.senat.fr/rap/l02-215-1/l02-215-1.html
Mistry R (2010) Water becomes a legal human right. http://civicusalpha.org/ehhr3.0/component/content/article/9-human-rights-news/108-water-becomes-a-legal-human-right
Möllers C (2005) Gewaltengliederung. Legitimation und Dogmatik im nationalen und internationalen Rechtsvergleich. Mohr Siebeck, Tübingen
Müller T (2011) Belgien – handlungsunfähig in der Krise. http://www.zeit.de/wirtschaft/2011-11/belgien-euro-krise
Müller JP, Cottier T (1995) Estoppel. In: Bernhardt R (ed) Encyclopedia of public international law, vol II. Elsevier North Holland, Amsterdam, pp 116–119
Muralidhar S (2004) Economic, social & cultural rights: an Indian response to the justiciability debate. In: Ghai Y, Cottrell J (eds) Economic, social & cultural rights in practice: the role of judges in implementing economic, social & cultural rights. Interrights, London, pp 23–32
Muralidhar S (2006) The right to water: an overview of the Indian legal regime. In: Riedel E, Rothen P (eds) The human right to water. Berliner Wissenschafts-Verlag, Berlin, pp 65–81
Murty MN, Kumar S (2011) Water pollution in India: an economic appraisal. In: Infrastructure Development Finance Company (ed) India infrastructure report 2011 – water: policy and performance for sustainable development. Oxford University Press, New Delhi, pp 285–298
Musil A, Kirchner S (2012) Das Recht der Berliner Verwaltung, 3rd edn. Springer, Berlin/Heidelberg
Nettesheim M (2012) Commentary on Art 11 EUT. In: Grabitz E, Hilf M, Nettesheim M (eds) Das Recht der Europäischen Union. C H Beck, Munich
Organisation for Economic Co-operation and Development (OECD) (2006) Liberalisation and universal access to basic services. OECD, Paris
Organisation for Economic Co-operation and Development (OECD) (2007) Environmental performance reviews: Belgium. http://www.oecd.org/env/country-reviews/environmentalperformancereviewsbelgium2007.htm
Office of the High Commissioner for Human Rights (OHCHR) (2005) Economic, social and cultural rights: handbook for national human rights institutions. United Nations, New York/Geneva
Ouvry B (2007) Access to water as a human right, different ways to fulfill this vital commitment: the cases of Belgium and its regions: speaking note on the occasion of the consultation on human rights and access to safe drinking water and sanitation. Office of the High Commissioner for Human Rights, Geneva, 11 May 2007. http://www.unece.org/env/water/whmop1/roundtable/Ouvry.pdf
Pant R (2003) From communities' hands to MNCs' Boots: a case study from India on the right to water. http://www.righttowater.info/wp-content/uploads/india_cs.pdf

Papier HJ (2012) Commentary on Article 14. In: Maunz T, Dürig G et al (eds) Grundgesetz-Kommentar, Band I. C H Beck, Munich
Parienté J (2012) TVA de droite, TVA de gauche: le jeu des différences selon Hollande. http://www.lemonde.fr/politique/article/2012/11/14/tva-de-droite-tva-de-gauche-le-jeu-des-differences-selon-hollande_1790190_823448.html
Partzsch L (2006) Wasser in der Krise – Das Beispiel Südafrika. Solidarische Welt 196(4):4–6
Powlowski M (2006) Making public health motivated evictions consistent with the right to housing. Quinnipiac Health Law J 9(2):271–318
Püttner G (2002) Das grundlegende Konzept der Daseinsvorsorge, Kommunale Daseinsvorsorge: Begriff, Geschichte, Inhalte. In: Hrbek R, Nettesheim M (eds) Europäische Union und mitgliedsstaatliche Daseinsvorsorge. Nomos, Baden-Baden, pp 32–38
Riedel E (1986) Theorie der Menschenrechtsstandards. Duncker & Humblot, Berlin
Riedel E (1991) Standards and sources – farewell to the exclusivity of the sources triad in international law? Eur J Int Law 2(2):58–84
Riedel E (2005) The human right to water. In: Dicke K et al (eds) Weltinnenrecht – Liber amicorum Jost Delbrück. Duncker & Humblot, Berlin, pp 586–606
Riedel E (2006) The human right to water and general comment No. 15 of the CESCR. In: Riedel E, Rothen P (eds) The human right to water. Berliner Wissenschafts-Verlag, Berlin, pp 19–36
Riedel E (2011) International covenant on economic, social and cultural rights (1966). In: Wolfrum R (ed) Max Planck encyclopedia of public international law. Oxford University Press, Oxford (online edition)
Ringwald R (2008) Daseinsvorsorge als Rechtsbegriff: Forsthoff, Grundgesetz und Grundversorgung. Peter Lang, Frankfurt am Main
Roberts A (2001) Traditional and modern approaches to customary international law: a reconciliation. Am J Int Law 95(4):757–791
Rosencranz A, Rustomjee S (1995) Citizens' right to a healthful environment in India. Environ Policy Law 25(6):324–328
Rudolf B (2007) Menschenrecht Wasser: Herleitung, Inhalt, Bedeutung, Probleme. In: Rudolf B (ed) Menschenrecht Wasser? Peter Lang, Berlin, pp 15–43
Rüfner W (2006) Daseinsvorsorge und soziale Sicherheit. In: Isensee J, Kirchhof P (eds) Handbuch des Staatsrechts der Bundesrepublik Deutschland, vol IV, Aufgaben des Staates, 3rd edn. C F Müller, Heidelberg
Ruge R (2003) Das Grünbuch der EG Kommission zu Leistungen der Daseinsvorsorge: Neue Konturen für einen alten Irrwisch? Zeitschrift Rechtspolitik 36(10):353–357
Santos C, Villareal A (2006) Uruguay: direct democracy in defense of the right to water (revised version). http://www.tni.org/sites/www.tni.org/archives/books/wateruruguayrev.pdf
Scanlon J, Cassar A, Nemes N (2004) Water as a human right? International Union for Conservation of Nature and Natural Resources (IUCN) Environmental Policy and Law Paper No. 51. Gland/Cambridge
Schachter O (1991) International law in theory and practice. Nijhoff, Dordrecht
Schwarzenberger G (1947) The inductive approach to international law. Harv Law Rev 60(4):539–570
Schwarzenberger G (1965) The inductive approach to international law. Stevens, London
Scott C, Alston P (2000) Adjudicating constitutional priorities in a transnational context: a comment on Soobramoney's legacy and Grootboom's promise. South Afr J Hum Rights 16(2):206–268
Seewald O (2006) Kommunalrecht. In: Steiner U (ed) Besonderes Verwaltungsrecht, 8th edn. C F Müller, Heidelberg, pp 1–170
Shah P (1997) International human rights: a perspective from India. Fordham Int Law J 21(1):24–44
Shaw MN (2008) International law, 6th edn. Cambridge University Press, Cambridge
Shaw A, Gotora G (2013) Zimbabweans vote in favor of new constitution – Zimbabwe: 94 percent vote for new constitution, but police jail rights lawyer for 3rd night. http://readingeagle.com/article.aspx?id=459940

Shelton D (2002) Human rights and the environment: jurisprudence of human rights bodies. Environ Policy Law 32(3–4):158–167

Simma B, Alston P (1988–1989) The sources of human rights law: custom, jus cogens, and general principles. Aust Yearb Int Law 12:82–108

Smets H (2005) La solidarité pour l'eau potable. Research Paper Series of the Academie de l'eau, Paris. http://www.academie-eau.org/admin/fichier_publication/70-2005_Pour_un_droit_effectif_a_l_eau_potable_Avril_2005-1271951606fichier_publication0.pdf

Smets H (2006a) The right to water at the 4th world water forum in Mexico. http://www.righttowater.info/wp-content/uploads/The-Right-to-Water-at-the-World-Water-Forum-Mexico-2006.pdf

Smets H (2006b) The right to water in national legislations. Research paper series of the Agence Française de Développement, Paris. http://www.emwis.net/documents/database/semide/PDF/right-to-water_EN

Smets H (2008a) De l'eau potable a un prix abordable: La pratique des États. http://www.academie-eau.org/admin/fichier_article/65-2008_De_l_eau_potable_a_un_prix_abordable-1269361608fichier00.pdf

Smets H (2008b) Rights and duties associated with the right to water. In: Fischer-Lescano A et al (eds) Frieden und Freiheit, Festschrift für Michael Bothe zum 70. Geburtstag. Nomos, Baden-Baden, pp 711–750

Smets H (2012) Charging the poor for drinking water: the experience of continental European countries concerning the supply of drinking water to poor users. Paper presented at a workshop hosted by publicpolicy.ie on June 19 2012. http://www.publicpolicy.ie/wp-content/uploads/Water-for-Poor-People-Lessons-from-France-Belgium.pdf

Stein T (1988) Remarks on custom. In: Cassese A, Weiler JHH (eds) Change and stability in international law making. Walter de Gruyter, Berlin, pp 12–14

Szabo A (2010) The value of free water: analyzing South Africa's free basic water policy. http://www.uh.edu/~aszabo2/water_szabo_0512.pdf

Tomuschat C (2004) Human rights: between idealism and realism. Oxford University Press, Oxford/New York

Tully S (2005) A human right to access water? A critique of general comment No. 15. Neth Q Hum Rights 23(1):35–63

Tully S (2006) Flighty purposes and deeds: a rejoinder to Malcolm Langford. Neth Q Hum Rights 24(3):461–472

United Nations Educational, Scientific and Cultural Organization (UNESCO) (2003) The United Nations world water development report: water for people – water for life. http://unesdoc.unesco.org/images/0012/001297/129726e.pdf

Unknown (2007) What price for the priceless? Implementing the right to water. Harv Law Rev 120 (4):1067–1088

Vedder C (2012) Commentary on article 11 TEU. In: Vedder C, Heintschel von Heinegg W (eds) Europäisches Unionsrecht: Handkommentar. Nomos, Baden-Baden

Weisburd AA (1988) Customary international law: the problem of treaties. Va J Transnatl Law 21 (1):1–46

World Health Organization (WHO) (2012) Glass 2012 Report: UN-water global analysis and assessment of sanitation and drinking water – the challenge of extending and sustaining services. http://whqlibdoc.who.int/publications/2012/9789241503365_eng.pdf

Winkler IT (2010) Respect, protect, fulfil: the implementation of the human right to water in South Africa. In: Cullet P et al (eds) Water governance in motion: towards socially and environmentally sustainable water laws. Foundation Books, New Delhi, pp 415–444

Winkler IT (2012) The human right to water: significance, legal status and implications for water allocation. Hart, Oxford

Wolff HJ et al (2007) Verwaltungsrecht I, 12th edn. C H Beck, Munich

Young K (2008) The minimum core of economic and social rights: a concept in search of content. Yale J Int Law 33(1):113–175

Chapter 3
Philosophical and Conceptual Approaches to a Human Right to Water

As we have established in Chap. 2, there is not easily a comprehensive, legally binding and at the same time self-standing human right to water in international law. Only if one accepts a derivative right as a self-standing right, one can conclude that the right to water exists in international law, namely as a right of its very own kind or with a unique status. The right also arguably exists by now in customary law (albeit with a so far relative weak fundament in State practice).

The next logical step forward is to follow two basic questions: Firstly, under which conditions can there be such a thing as right to water (and if so, what conceptual reasons held it back for so long?); and secondly, should there be one at all? The purpose of this chapter is thus not only to undertake the often forgotten inquiry of whether such a right can exist in a logically coherent manner[1]; it also asks in what way to re-conceptualize existing approaches to best reflect such a right, so it has solid normative foundations.

The reason to include a theoretical part is the attempt to examine the concept of a human right to water from the very beginning. One should take for granted as little as possible. The fact that a human right to water is increasingly assumed to be existent should not mislead a scientific inquiry to any hasty conclusions. Already the bare assumption that a right to water would even be possible under the common legal terms and categories, should be proven first, before accepted. This chapter will discuss these possible objections, even those that perhaps appear improbable at first. It is only in this way that one can make a convincing case *for* a right to water, if it shown, that a convincing case *against* it, cannot be made.

For this purpose we shall use a method which Descartes called "the method of doubt".[2] This is the idea that a claim, as the claim in favor of a human right to water,

*All cited webpages were last accessed on 1 August 2013.

[1] Some of the few authors who have recently raised this question are Bleisch (2006) and Ladwig (2007); partly also Veigha de Cunha (2009).

[2] See Descartes (1999). On this method, see for instance Broughton (2002); and Curley (1999).

P. Thielbörger, *The Right(s) to Water*,
DOI 10.1007/978-3-642-33908-0_3, © Springer-Verlag Berlin Heidelberg 2014

can only convincingly stand, if every possible logical objection to it can be adequately disproven.

We can identify the three major theoretical challenges already in the three elements of very term "human right to water" itself. Firstly, is *water* a possible content of a right? Secondly, is a suitable legal approach to the basic human need for water a *right*? And thirdly, would this right be correctly qualified as a *human* right? These three elements will provide for the first part of the structure of this chapter. In a second part, this chapter poses the question how a coherent and holistic approach to a human right to water would have to look like, and how it would relate to the recognitions given to the right in domestic legal orders.

3.1 The "Water" Element of the Term "Human Right to Water"

It appears that a human right to water would somehow be different from more traditional rights, such as the right to express an own opinion, the right not to be deprived of one's own life or to freely choose a religion. What is it in water as a human right that accounts for this difference?

3.1.1 Objection: We Cannot Hold a Right to a Tangible Thing

At first sight, it is the content of the right: water. Water is a tangible thing. We can take hold of it. Its overall amount is limited. These facts are different from the content of many traditional rights. The contents of most traditional rights have certain defining characteristics in common which water does not share. Traditional rights are more about abstract values: the right to have an own voice, to maintain physical integrity or to participate in a religious community. They are usually not about tangible and quantifiable things. Should this difference hold us back from accepting a human right to water?

There is no rule or reason that prevents tangible things from forming the content of rights. Firstly, some other more modern rights in fact also concern tangible things, such as the right to housing or the right to food (albeit both are part of the abstract value that is an adequate standard of living). Both are accepted as human rights—so why should water not be a possible content of a right? Secondly, even more traditional rights are often factually connected to tangible objects. The law simply chooses to express the relation between tangible things and the right-holders in abstract terms. To take just one example, private law does usually not say: X has a right to a car. He has either the right to own a car ("property"), to drive with it ("right to use") or to buy the car in the future ("purchase option"). In this sense, the content of the human right to water is not in fact the object "water" itself. It is more a right concerning the relation between the object of water and the rights-holder.

It entitles us to water-related actions that we should be enabled to carry out, just as with any other right. This action can be, for instance, to drink or to clean oneself. It is these actions that are crucial to the right.

One might even argue that, if the right is connected to a substance, the use of this substance is usually more important in the recognition of the right than the substance itself. If we look for instance at the discussion on the right to fresh air, an upcoming topic in environmental law,[3] it is a discussion on a right to unimpaired and healthy breathing more than on a right to a substance made of oxygen, nitrogen and carbon dioxide. The term human right to water is therefore misleading. It is just a convenient and shorter version of a more complex right, a right to privilege certain uses of water over others. One lesson is then to clearly distinguish what uses the right to water is intended for—and to prioritize the most essential actions (drinking; washing; cooking) from less essential ones.[4]

For these reasons, the general objection that water is an unsuitable content of a right in the first place cannot be upheld.

3.1.2 *Objection: The Time to Accept Water as a Right Is Not "Now"*

This leads to a second challenge of the right. *When* is the right timing for water to become a human right? Is it too early? Is it too late? A possible objection against the human right to water is that "*now*" is not the right time to acknowledge it.[5] This argument is based on the notion that if a resource is almost extinct, it makes no sense to accept it—its fulfilment is no longer possible. For limited resources like drinking water, this means there is a time when it is too *late* to accept them as forming the content of a right. On the other hand, perhaps the very moment when we begin to "take rights seriously" is exactly this period in which we most need them. Ronald Dworkin has famously used such a claim to argue that rights necessitate some restrictions on democracy. To "have a point" they must "trump" political goals that seek to maximize the overall political welfare ("utilitarianism").[6] To the extent that our concept of political justice is based on some idea of popular democratic rule, the operation of human rights may entail sacrifices for democratic rule, even if at the same time they are a necessary pre-condition for the

[3] See for instance the contested "Feinstaub"-judgment of the BVerwG, Case No. 7 C 36/07, 27 September 2007, BVerwGE 129, 296.

[4] See the suggestion for a framework for prioritization of different uses of water: Winkler (2012), pp. 150 ff.

[5] In fact, some of Tully's previously discussed arguments (Tully (2005); *id.* (2006)) go in this direction.

[6] Dworkin (1984), pp. 153–159.

creation of a democratic society.[7] The demand for a right must therefore be pressing enough to justify these societal costs. Without the pressing need for a right, there is simply no driving force for its political discussion and creation. This is exactly shown by the above mentioned right to clean air: only with the increasing pollution of the air has the threat to human health become sufficiently pressing for a serious discussion of this right to take place.

It is exactly in this phase between abundance and extinction when we discuss the rise of a new right. This phase, which one could call progressing scarcity, is the only appropriate time for the recognition of a right.[8] Whether this time is "now" in the case of water, is ultimately for scientists to find out. However, the fact that water is getting scarcer and scarcer[9] and predictions of global water shortages are becoming more prominent, gives some indication that the current status of fresh water could be exactly in that mentioned timeframe of progressing scarcity.

3.1.3 Conclusion

The overall answer to the question of whether "water" is a suitable content for a right is positive, but with restrictions. That water is a tangible thing is no valid counter-argument, as other rights include tangible objects as well; it is more an (acceptable) inaccuracy of the term itself. Secondly, only in a certain timeframe, namely between abundance and extinction, the evolution of a new right to a resource is feasible. The scientific findings available to us indicate that—far from being too early and hopefully not being too late—the correct timing for the rise of a human right to water might be exactly now.

3.2 The "Right" Element of the Term "Human Right to Water"

Why should one address the undeniable human need for water by the means or mechanisms of a "right"? Is addressing water needs through the discourse on a right a logical and coherent endeavor at all?

---

[7] *Ibid.*, p. 155.

[8] As Benjamin Franklin once phrased it: "only if the well is dry, we know the worth of water" (Franklin (1793)).

[9] For an overview, see for instance WHO and UNICEF (2006); see also UNESCO (2012), especially pp. 315 ff.

3.2.1 The Concept of a Right

To address this question, we first have to better understand roughly what the term "right" means. The struggle to find a definition of the term is probably as old as the term itself and could not possibly be adequately summarized at length here. However, at least one should be aware of the common distinction of *being* right and *having* a right. In this sense, the term "right" has two senses, a moral one and a political–legal one: rectitude and legal entitlement.[10] A rights-holder cannot, or only under very strict conditions, be prevented from using his or her right, even if it might be morally "wrong" for him or her to so use it. In the contrary sense, even if something appears morally justified, this does not necessarily give it the status of a legal right. In other words, if we discuss rights, we do not only discuss abstract values such as liberty and equality, we discuss the concrete entitlements to realize these values.[11]

With this clarification in mind, one can think of possible objections to the idea of having a *right* to water. The first question that arises is a rather basic one: as water is absolutely necessary to sustain human life and cannot be substituted, can there be a right, where there is no *choice*? Can we have a right to something which we simply must have and cannot meaningfully live without?

3.2.2 Objection: We Cannot Accept a Right Without Choice

Physically speaking, there is, of course, always choice. Individuals can decide not to make use of their right to water, and consequently abstain from drinking.[12] This is not an unrealistic scenario, if we think of the hunger-strikes of political prisoners. Admittedly, this choice will, if continued, ultimately lead to death within a few days.

One could argue that, more generally speaking, the basic idea about rights is that they represent freedoms or liberties. H.L.A. Hart[13] argued famously that it is the right to liberty that all other rights depend on. We have a right, but whether we make use of it or not, must be our choice. This is a part of *Hart's* famous "will theory" that rights are a way of avoiding paternalism: they put the rights-holder "in the driving seat" as to whether they should use their entitlement or not.[14] A right to

---

[10] Out of many: Donnelly (2003), p. 602.

[11] *Ibid.*, p. 603.

[12] For that element of a right to water that is connected to personal hygiene this problem is not relevant: rights-holders can decide on whether to wash themselves, or not, etc.

[13] Hart (1954), p. 49. See, for other contemporary defenses of will theory: Wellman (1995); Graham (1996).

[14] Hart (1954), p. 49.

do X is only a complete right, if the right *not* to do X is also part of the right, so the argument goes. So, can one accept a right, if we cannot surrender or renounce it?

In practice, we accept a number of rights that do not leave us with a choice. In some countries, for instance in Argentina, Belgium and Australia, voting is a right, even if the exercise of the right is at the same time a duty ("compulsory vote").[15] Its non-exercise is illegal and in some countries leads to sanction. A second example for a right without choice would be the right to a fair trial: we would never allow—at least in a State governed by the rule of law—that the accused renounces or rescinds the right to a fair trial, thereby encouraging his or her judge to make an arbitrary decision. And finally, if a choice was a necessary condition for accepting a right, rights-holders without the intellectual capacity to make a well-reasoned choice about the exercise or the non-exercise of a right (for instance children or lunatics) could not have rights[16]—a result we certainly would not want to support. These examples show that the possibility of surrender or permanent non-exercise of a right is not a necessary component of a right. If we refer to *MacCormick's* famous dilemma on children's rights[17]—that people (such as vulnerable children) who need rights the most would be precisely the ones, under will theory, who cannot claim or enforce them—one could even make the argument that the rights that do not involve any choice are often some of the most central to human rights' very architecture.

Thus, the argument that a right to water would be a right whose exercise or fulfillment we cannot realistically give up is significant yet not fatal to accepting water as the subject of a right in the first place.

3.2.3 *Objection: There Is No Added Value of a Right to Water*

Another related question is whether a rights-based approach is the best answer to the problem of a global water-crisis. What distinguishes a rights-based approach from other approaches? What is the added value of a right? Is a *right* to water not simply a redundancy,[18] if we assume the general duty of the State to supply its population with the most essential goods? Perhaps the individual and societal need for water is better addressed not through rights but through the general political duty of governments to act for the general welfare of their citizens?

There are three arguments that this objection neglects. Firstly, the concept of rights serves to illustrate the importance of respect for individuals in legal regimes,

[15] For an overview of countries with compulsory vote, its history and its legal enforcement, see IDEA (2012).

[16] See for example Arnold (1978), p. 80. It is, in essence, the well-known counter-argument of the interest theory against the will theory.

[17] MacCormick (1976).

[18] See for instance Arnold (1978), pp. 77 ff. Arnold argues in favor of such a redundancy.

by allowing them to make autonomous claims in a way which is not inherent in the idea of a State's duty.[19] It is this respect for the human being—a respect that arguably derives from the very idea of human dignity—that ultimately makes a rights-approach preferable to any other one.

Secondly, it has already been pointed out by H.L.A. *Hart* that rights usually entail duties, but that the converse argument does not always hold *per se*.[20] He argued that there are duties that do not give rise to a corresponding right. This theory is also backed up in practice by judicial decisions. One of these examples we saw already in Chap. 2 by analyzing the German legal regime: the German Constitutional Court ("Bundesverfassungsgericht") has repeatedly decided that a citizen's entitlement to an existential minimum ("Existenzminimum") does not automatically derive from a State's duty to provide basic general services ("Daseinsvorsorge"), but needs to be derived from additional individual constitutional rights.[21] The court thereby recognized a duty of the State which does not automatically give rise to a corresponding right for the individual. Thus, arguing that existent State duties make an individual entitlement redundant is not convincing. Similarly, Philip Alston's suggestion not conjure up ever new rights for each and everything, but to apply a certain quality control for new rights,[22] is also not a convincing argument against a right to water. It is generally true that adding new rights might devalue the currency of human rights. However, given the utmost importance of water for human survival, Alston's critique (as deserved as it is in general terms) does not apply to a human right to water.

3.2.4 Objection: There Cannot Be a Right, if it Cannot Be Enforced

One can argue that a human right, if credible, demands the possibility to enforce the claims that derive from it.[23] This objection has in fact two facets.

Firstly, we cannot claim a right, if this is not mirrored on the other hand by institutions to promote and monitor the realization of these rights. Onora O'Neil voiced this critique by saying that the content of economic and social rights and the relation between right- and duty-holder often remains "wholly obscure", claiming that economic and social rights "*must be institutionalized: if they are not, there*

---

[19] A similar argument is put forward by Wasserstrom (1964).

[20] Hart (1973), pp. 171–201.

[21] See BVerfG, Case No. 1 BvR 220/51, 19 December 1951, BVerfGE 1, 97, 104; BVerfG, Case No. 1 BvL 4/74, 18 June 1975, BVerfGE 40, 121, 133; BVerfG, Case No. 1 BvL 14/76, 21 June 1977, BVerfGE 45, 187, 228. On this issue in general, see Friauf (1971), p. 676.

[22] Alston (1984).

[23] Again, some of Tully's arguments against General Comment No. 15 (Tully (2005); *id.* (2006)) go in this direction.

is no right."[24] Secondly, so it is argued, socio-economic rights, like the right to water, are not judicially enforceable by courts.[25] Given how unspecific their content is,[26] some authors have even argued that they are too indeterminate to be properly enforced in the first place (no matter by which institution)[27]: they should not be a matter of enforceable law at all.[28]

With respect to the lack of international institutions, one must counter these arguments by emphasizing that establishing these rights is exactly the necessary incentive to create such institutions. Why should one create institutions to promote and monitor rights before they are even accepted? One needs to start with either rights or institutions, or else one never starts at all, creating a vicious circle in which progress towards the realization of new rights is permanently halted. However, one can also counter this argument on practical grounds: the challenge of lacking institutionalization of economic and social rights is becoming less and less relevant. In 2008, the UN General Assembly (UNGA) decided to introduce an Optional Protocol to the Covenant on Economic, Social and Cultural Rights,[29] aiming to give them in the future a similar status as civil and political rights in terms of their enforceability. Once in force, the Optional Protocol will establish an international individual complaint and inquiry mechanism.[30] As we will see in greater detail in Chap. 4, important monitoring institutions with respect to access to water have recently been developed, foremost the Special Rapporteur on the human right water and sanitation. As a result, the international institutions that are deemed necessary as the logical corollary of a right have developed in the case of water or are at least currently developing.[31] This objection seems to have been overcome.

With regard to socio-economic rights' content being too indeterminate to be enforced, this is a concern about all human rights, albeit maybe for socio-economic rights in particular. However, one should not overlook the significant contribution of the Committee's General Comments[32] and the multitude of additional literature[33] on the International Covenant on Economic, Social and Cultural Rights (ICESCR) which have largely helped to make the content of socio-economic rights more concrete. They have thereby become much more enforceable than used to be the case in the past.[34]

[24] O'Neill (1996), pp. 131–132.

[25] Vierdag (1978), p. 76; also see An-Na'im (2004), p. 13.

[26] Klee (2000), p. 97.

[27] Cranston (1967), p. 47; Vierdag (1978), p. 93.

[28] Cranston (1967), p. 47.

[29] UNGA, *Optional Protocol to the International Covenant on Economic, Social and Cultural Rights*, A/RES/63/117, 5 March 2009.

[30] *Ibid.*, Art. 2 ff. (called "Communications") and Art. 11 (called "Inquiry Procedure").

[31] Winkler (2012), p. 106.

[32] Coomans (2002), p. 194; Winkler (2012), pp. 104–105.

[33] Sepúlveda (2003), pp. 132 ff.

[34] Winkler (2012), p. 105.

3.2.5 *Objection: We Cannot Accept Rights Whose Realization Is "Impossible"*

Can we accept a right whose realization would ultimately be impossible, given that it is aimed towards a truly global population? Accepting water as a human right would ultimately mean the whole global population should receive access to water. Yet such a right for all people is impossible to realize, at least immediately. Whether the right to water would be such an "impossible" right, is ultimately a question for scientists.[35] However, for the purposes of this study, one should assume the worst case in terms of the argument—that there is not enough water for all humans.[36]

This argument is not new: the argument that one cannot have a right to the impossible has often been used as one of the main arguments against the use and development of social and economic rights in a general sense.[37] As Maurice Cranston claims, it is "totally impossible" to translate socio-economic rights in the same way into positive enforced legal rights by political and legal action[38]; claiming welfare rights in poor countries, where industrialization has hardly begun, would be "vain and idle".[39] Assuming ever more of these new rights, so the argument goes, adds to the incoherence of human rights theory and hinders the effective protection of genuine human rights.[40]

Three arguments, however, stand in the way of this argument. Firstly, legal rights have a meaning beyond a concrete legal claim. Legal rights are at the same time general expressions of the values accepted in a legal order. These values have to be taken into account in decision-making processes, both political and judicial. Hence, legal rights can hold a value, even if their main content for the right holder might be impossible to realize.[41]

Secondly, as famously Henry Shue pointed out, some social and economic rights (which he calls subsistence rights) pass the test of practicality as easily as some civil and political ones (which he calls security rights).[42] It is a false belief that civil and political rights ask less of the State (only negative action), while assuming that

---

[35] See UNDP (2006).

[36] If we assumed to the contrary that there was enough potable water for all, the problem would not even arise or at least to a much lesser degree.

[37] Vierdag (1978), p. 69; van Hoof (1984), p. 98; Macfarlane (1985), p. 9.

[38] Cranston (1967), p. 166.

[39] *Ibid.*, p. 50.

[40] *Ibid.*, p. 164.

[41] Some authors call this the distinction between "direct" and "indirect" obligations arising from rights, see for instance Sen (2004), p. 315. Thus, a right which creates an impossible direct obligation can still create a valid impossible obligation.

[42] Shue (1980), pp. 185–186.

socio-economic rights always ask for more (e.g. positive State action). The fulfillment of socio-economic rights is quite possible, so Shue, at least over time.[43]

Thirdly, it is problematic to allow a State too easily to call a right "impossible". If guaranteeing the protection of rights gives the State part of its very legitimacy, a State cannot easily dismiss the acceptance of essential rights that are crucial to human survival. It therefore appears plausible that States are bound to make considerable efforts to enable themselves to guarantee basic rights where they are most needed. This may be even more so in the case of rights that are particularly hard to realize. Here, the dilemma pointed out by Neil MacCormick is turned around: States have to try and make impossible rights as possible as possible, especially those as indispensable to life as a right to water would be. It follows from the State's protective function that the more essential a right in concrete circumstances, the *closer* to our grasp the State has to bring its realization. As such, States will eventually have to think of new means that go beyond daily politics in order to guarantee these essential rights. The internal re-organization of the State, greater international cooperation or investment in new technologies such as desalination may all be required for this purpose. Even if we accept that rights must have an element of "practicability" to be rights, this practicability does not only include what a State can offer and afford at present and under the current organizational circumstances. It goes much beyond this.[44]

3.2.6 Conclusion

The overall answer to the question of whether a "right" can be an appropriate legal answer to the human need for water can therefore be answered positively. Although water is a right without choice, it gives an added value for the individual's moral position. The argument that water is limited and that the right to water under the current circumstances might be considered an (almost) impossible right, imposing currently substantial duties, is valid, but only in parts. States derive their legitimacy *inter alia* from their duty to protect and promote the rights of their people. In turn, they are under a duty to ensure that the realization of basic rights does not *become* impossible. They have to make enormous efforts to protect and promote scarce resources that are essential for human survival. The very fact that a human right to water is precarious or difficult to achieve cannot lead us to abandon the concept as a whole. Instead, these are precisely the circumstances in which the rights-discourse becomes most important. Keeping these highlighted particularities in mind, there are good reasons to phrase the issue of access to water in human right terms.

---

[43] *Ibid.*, pp. 185–186.

[44] In a similar sense: Shue (1980), p. 16. If private persons must make such arrangements, States will have to do so even more.

3.3 The "Human" Element of the Term "Human Right to Water"

Finally, we must approach the third element of the term in question: can such a right to water be qualified as a *human* right?

3.3.1 The Identifiable Schools of Human Rights[45]

One should start by mentioning that there is a long-standing tradition of doubting that *human* rights, or natural rights, exist at all. Jeremy Bentham famously called the very idea "nonsense upon stilts",[46] claiming that a "*right, the substantive right, is the child of law; from real laws come real rights; but from imaginary laws, from law of nature, come only imaginary rights.*"[47] This cynical view of human rights is the expression of a comprehensive dismissal of human rights and aims to refute any belief that people have in unconditional rights, rights that we hold just by the virtue of being human. Alistair MacIntyre has more recently revived Bentham's critique of human rights, claiming that "*there are no such rights, and belief in them is one with belief in witches and unicorns*".[48] There is just no evidence for their existence, so McIntyre's claim goes. According to these critical scholars, if the ongoing discourse around human rights can hold any value at all, it is largely as a policy-making tactic—used to support people in need—rather than as a concrete legal claim.[49] This general denial of human rights is a very particular view, today only supported by few scholars. For reasons of space, it cannot be fully discussed here. In this chapter, we shall rather follow the largely prevailing opinion—that a sub-category of rights exists that we can rightly call human rights—and focus on the question of how these rights are defined, before applying the different definitions in answering the question whether a right to water is one of these rights.

In order to answer the question whether a right to water must necessarily, or can potentially, be qualified as a *human* right, we must first establish an understanding of what a human right is. Of course, there is little space here to review the entire literature on the question of the philosophic foundations of human rights.[50] We shall rather identify the main streams in present understandings of human rights,

---

[45] This book follows the very useful categorization of human rights suggested by M. B. Dembour. This recent approach captures appropriately a framework to classify a variety of very different human rights; see Dembour (2010), pp. 1–20; *id.* (2006), pp. 232 ff.

[46] Bentham (1843), p. 501.

[47] *Ibid.*, p. 523.

[48] McIntyre (2007), pp. 68–69.

[49] See for instance, Brown (2004), pp. 461–462; and Speed (2008), p. 181.

[50] For a longer discussion of the sources of human rights, see, out of many, Shestack (1998) and the aforementioned Dembour (2010), pp. 1–20.

assessing whether a right to water would qualify as a human right under these different streams.

The oldest approach that justifies the existence of human rights lies in religion. At first sight, this religious approach might appear paradoxical, as many practices proscribed by world religions today appear to be in conflict, if not incompatible, with modern understandings of human rights, such as equality or individual freedoms (e.g. the proclaimed inferiority of women in some religions). At the same time, a number of religious beliefs and concepts underlie our modern conception of human rights. A dominant idea in Jewish and Christian scripture is that man was created as the "image of god".[51] The Qu'ran and the Hindu Bhagavad-Gita holds similar assumptions. Humans, in the view of many of the major religions, hold a high value of worth and dignity: because humans were given their rights from a divine source, these rights in turn are inalienable by mortal authority.

A similar traditional approach is that of natural law scholars. They conceive human rights as "given". Humans hold these rights just by virtue of their natural existence. As Jack Donnelly puts it, human rights are "*the rights one has simply because one is a human being*", thus they are held universally by all human beings".[52] The source of human rights is then man's moral nature.[53] Similarly, in the words of Michael Perry, the concept of human rights is "*the idea that there is something about each and every human being, simply as a human being*" and that certain things must not be done to any human being and certain other things ought to be done for every one of them.[54]

Opposed to this natural or religious school of theory, there are two schools that assume that human rights are rights made by man. What one could call a deliberative stream describes human rights as those rights that have been agreed upon as such by humankind. For the German philosopher Jürgen Habermas, citizens are those individuals who deliberate, acting as a constitutional assembly. They decide upon the rights they have. In this sense, the scope of their public autonomy is not restricted by any natural or moral right just waiting to be put in place.[55] Similarly, Michael Ignatieff claims that human rights are no absolute values, but rather a language that is the basis for all deliberation.[56] The fundamental moral commitment entailed by rights is then not to respect, it is to deliberation or the ability of communities to frame basic entitlements through a free and open deliberative process.[57] The nature and extent of the rights humans hold is just a social fact[58];

---

[51] See in the Bible, English Standard Version, Genesis 1:27, "So God created man in his own image, in the image of God he created him; male and female he created them."

[52] See Donnelly (2003), p. 1.

[53] *Ibid.*, p. 14.

[54] Perry (1998), p. 13.

[55] Habermas (1996), p. 127.

[56] Ignatieff (2001), p. 95.

[57] *Ibid.*, p. 84.

[58] Campbell (2006), p. xii.

in this way, human rights are political and societal values from which societies have drawn.

Related to, but slightly different, from this understanding, is the idea that human rights are rights over which humankind has fought conflicts. This school links human rights closely to the social movements that fought for them. This assumes firstly that human rights hold a particular importance (who would fight for the irrelevant?) and that it is difficult, or costly, to realize them (why would one otherwise object to their creation?). For Upendra Baxi, human rights' main task is to voice human suffering and make it visible, being the first to ameliorate it.[59] Similar, for Neil Stammers, social movements (as the groups that engage in these social struggles) are the key originating source of human rights.[60] The emergence of human rights is then the result of the human resistance and struggle against suppression and for the empowerment of the suppressed.

3.3.2 *Water as a Human Right According to the Different Schools*

Having categorized the existing views on human rights, one must ask the question of whether a right to water qualifies as a human right according to these different views.[61]

As for the religious and natural schools, water is very closely linked to the nature of any human being. By virtue of their creation, be it divine or not, human beings need, more than almost anything else, access to safe drinking water every day. Independent of culture or political circumstances, the basic human need for water is undeniable and highly natural. In this sense, one could pointedly ask: what could qualify as a human right from a natural law perspective if not water? It is a right independent on the political will of a particular State or social movement: either all humans hold such a right, by virtue of their existence, or no human has it at all.

Under the deliberative schools following Habermas' doctrine, access to water would need to be a political value that societies have chosen to accept in a deliberative process. One might make the claim that this deliberative process has not yet come to an end, and that therefore access to water has not yet reached the fully-fledged status of a human right. However, it is impossible to deny that such a deliberative process on the right to water is currently ongoing. The question of water as a human right is greatly discussed, in politics and civil society alike, across most nations of the world. In the future, access to water will, with increasing clarity, be accepted as a human right by means of these deliberative processes. As this

[59] Baxi (2006), p. 2.

[60] Stammers (2009), pp. 1–2.

[61] One can leave out the obvious: the school which denies the existence of all human rights, would also dismiss a human right to water.

chapter is concerned with the general acceptability of access to water as a human right, rather than with the empirical question of whether it is already accepted, the fact that a deliberative process is ongoing indicates sufficiently that a right to water, if accepted, would easily qualify as a human right, if one follows a deliberative school of human rights.

Finally, for the last mentioned stream, human rights are those important rights for whose acceptance and realization mankind had to go through conflicts or war. Water has, since the early history of mankind, been one of the main causes for conflicts and wars. People, in particular indigenous people, often had to fight for their right to access water resources, which they had been using for generations. The history of water has been, and sadly might be in the future, linked to violence, armed conflict and power struggle. It has often been warned that water shortage is potential fuel for war, and is a likely cause of armed conflict in the future. Most recent examples include numerous societal unrests in which populations have violently objected to the privatization of water supply systems, fighting for their own "right to water".[62]

3.3.3 Conclusion

Thus, all streams would qualify a right to water, if so accepted, as a human right. It is naturally linked to human existence; societies are deliberating about its acceptance intensively, and people have been, and will be, prepared to face armed conflict over access to water. However, one should not misunderstand this finding. As we saw in Chap. 2, a right to water does not *need* to be designed as a human right only. There are many other recognitions of rights to water in national and regional legislation. This finding does not impede our conclusion here: that the right to water on the international level, if so accepted, would fulfill the criteria of a *human* right. The human right to water and recognitions of the right in domestic law complement each other; in which ways will be subject to further analysis in the following part of this chapter.

3.4 The Shape of a Right to Water: Combining Existing Approaches

We have so far established three things in this chapter: Firstly, water is generally a suitable content for a right. Secondly, there is an added value of tackling the human need to water through the means of a right. Thirdly, such a right can be qualified as a human right, whichever theoretical approach one takes to the term of human rights.

[62] See for instance Sect. 4.2.2.3 of this book.

Having disproven all conceptual obstacles to a right to water, this does not, however, say anything about the shape of such a right. A right to water could be accepted in many different ways. Answering this question involves distinguishing between a self-standing and a derived right, identifying its legal nature as a civil and political or as a socio-economic right, to define its content against other related rights, and to answer the question on which legal levels, domestic or national, a right to water should be recognized. We must also consider—if the right is recognized on different levels—how these levels relate to one another, if they clash or if what is required under a right to water on the different levels is significantly different. This chapter is, however, not a purely normative analysis. It rather aims to identify the best conceptual approach to a right to water which reflects the experiences and approaches seen in Chap. 2, and keeps in mind the theoretical challenges to the conception of the right, as raised in the previous three sections of this chapter.

3.4.1 A Self-Standing Right or a Derived Right?

Rights can either be self-standing (or independent) or they can be derived. Of course, even self-standing rights are often connected and interwoven with each other.[63] However, it makes a difference for the perception and potential of enforceability of a right whether it has its own origin or whether its legal source is founded in other rights. To understand the distinction, one must first understand the technique of legal derivation.

Rights can derive from another. A right which is grounded in at least one other right can be called a derivative right. Non-derivative rights to the contrary are sometimes called parent rights[64] or core rights.[65] There are theoretically three different ways in which rights can derive from parent rights. Firstly, a synthesis of several rights can add up to one new right. This procedure involves taking one aspect or the whole of each parent right and compounding these fragments to make

---

[63] Ultimately all human rights are "universal, indivisible, interdependent and interrelated", World Conference on Human Rights, *Vienna Declaration and Programme of Action*, A/CONF.157/23, 12 July 1993, par. 5. The question followed here is different: is a potentially new right distinct enough to be considered a right of its own or is it sufficient to understand it as part of another (already existent) right?

[64] This terminology of "Mutterrecht" ("parent right") is for instance suggested by Rudolf (2007), p. 26.

[65] This (much older) terminology is for instance suggested by Raz (1984), p. 197. However, the term "core right" is avoided in this book, as it might be easily confused with respect to the *minimum core* of a right, as developed in ESCR-Committee, *General Comment No. 3: The nature of States parties' obligations (Art. 2 par. 1)*, HRI/GEN/1/Rev.9 (Vol. I), 27 May 2008, at 7. Another thinkable terminology would be "primary" and "secondary" rights; this, however, can be easily misunderstood as primary and secondary law, and even more suggests a too strong hierarchy.

a new derivative right. Secondly, a derivative right can be derived from only one parent right and thereby replace it, if the phrasing of the old parent right is out-dated or no longer appropriate. Thirdly, a new right can also be derived from one single parent right, if that right is rather broad. In this case, the new derivative right is supposed to place further emphasis on one special aspect of the parent right.

Derivation is most often applied by courts (or quasi-courts like the Committee on Economic, Social and Cultural Rights (ESCR-Committee)), whose task it is to oversee and interpret existing legal rights. Only rarely do legislators revert to the technique of deriving rights from parent rights: they usually choose to generate new rights directly by legislative action. The reasons behind legal derivation are various. Judges and quasi-judges most often derive rights if they find a gap in the law without having the necessary legislative powers to bridge it. In this case they may be inclined to use their interpretive authority to create new rights by deriving them from parent rights. One logical consequence of derivation is the strong inter-linkage of the parent right and the derivative right: if a parent right is not commonly accepted, it is unlikely to create a commonly accepted derivative right; if a parent right lacks enforceability, the derivative right will also be difficult to enforce; and if a parent right is a negative right only, the derivative right can hardly be a positive one. Thus, parent and derivative rights share the same legal character.

General Comment No. 15, as has been discussed in Chap. 2, is ambiguous in whether it views the right to water as an independent right or as a derivative right. The release of a self-standing General Comment on the topic speaks in favor of a somehow independent right. The close normative connection to other rights of the ICESCR, as mentioned in General Comment No. 15, and the lacking power of the committee to create truly new rights, rather point in the direction of a derivative right. If the committee meant to create a derivative right, it still remains unclear which of the possible options of derivation the committee employed: either the right to water was derived from several rights[66] or the right to an adequate standard of living was meant to be the only parent right in question. This confusing situation between the right to water and other rights has rightly been criticized, most prominently by Amanda Cahill, who claimed that it is a right of a unique status.[67] How should one answer the question of the legal origin of the right to water in order to be most consistent with the right's normative content and with the variety of existing approaches in the different legal orders?

Firstly, one strong argument for assuming the independence of a right to water is to give it more visibility in the canon of all human rights, paying tribute to the fact that water is one of the most fundamental human needs of all. There is no reason, from a purely normative perspective, why rights that are less essential for survival,

[66] As was pointed out before, ESCR-Committee, *General Comment No. 15: The right to water (Art. 11 and 12)*, HRI/GEN/1/Rev.9 (Vol. I), 27 May 2008, at 97 (hereinafter "ESCR-Committee, *General Comment No. 15*"), in section "Genesis and Background of General Comment No. 15", also emphasizes the strong links to the right to health, the right to life and the notion of human dignity. In fact, the human right to water would then derive from up to four rights.

[67] Cahill (2005), pp. 394 ff.

such as the right to work, to take part in cultural life or to education, received the status of an independent right, as accepted under the ICESCR, while this status was denied to the human right to water. If independence of a right is in any way reflective of its importance, then this is a strong argument to assume the independence of the human right to water. Secondly, if a right to water was a derivative right, one could only assume a violation of that right, if at the same time the violation of the parent right(s) is proven.[68] This complicates the proof of a violation of the right to water significantly: it increases the argumentative effort that judges, or quasi-judges, need to undertake in order to assume a violation of the right to water. However, in practical terms, cases in which the right to water is violated, but the right to health or an adequate standard of living is not, are hardly conceivable.[69]

On the other hand, there are strong reasons why the right to water is a derivative right. Firstly, accepting a purely independent right to water would lead to the somehow strange outcome that it would receive a stronger recognition than the "including"-rights enumerated in Art. 11 ICESCR. It is one thing to argue that a right not explicitly included in the Covenant should get an equal status to the enumerated rights; to make the case for an even stronger self-standing recognition of an unmentioned right is quite a different thing. The second reason to see the right to water as a derivative right is the perception of the right amongst States: whenever mentioned on the international stage, its very close relations and connections with other rights are emphasized. Water is always referred to in connection with life, dignity, health or an adequate standard of living. Water is seen as one of the most important means to realize values like life, dignity and health. In order to justify the right to water, one needs to refer to values outside the right itself; thus, deriving the right from the rights corresponding with theses values, appears appropriate.

Thirdly, as Philip Alston has famously argued some 25 years ago, one should not conjure up new rights for each and every scenario; the creation of new rights calls for a certain quality control.[70] As Alston would most likely agree, a right to water, in its unique importance and concurrent fragility, would pass such a quality control. However, what lies behind Alston's argument is the effort to avoid doubling rights that are already existent. In other words: only if a right cannot be reflected as a combination of existing rights, only if a right brings something entirely new to the table, is there a need for creating an entirely new, independent right. Assuming that several rights, if read together, can entirely cover the content of the human right to water, there is no need for the creation of a new independent right.

Conceptualizing the human right to water as a derivative right, leads to the question what its parent rights are, in order to cover all facets of the right to water, and how the different parent rights would relate to each other.

[68] *Ibid.*, pp. 394–395; Rudolf (2007), p. 26.

[69] Rudolf (2007), p. 26.

[70] Alston (1984).

3.4.2 *Deriving from What? Plea for a More Integrated Approach*

As we have seen in Chap. 2, the prevailing opinion in international law assumes that General Comment No. 15 suggests the human right to water to fall exclusively under the category of economic, social and cultural rights. General Comment No. 15 is understood to name mainly Art. 11 (the right to an adequate standard of living) as the parent right for the right to water. However, as was also argued in Chap. 2, this is not the only way to interpret the General Comment. It also identifies strong linkages to other human rights, in particular the human right to health, the right to life and the notion of human dignity. We have also seen that the assumption of the right to water as a purely socio-economic right is not echoed by all national legislators and judges. Often the right is linked to civil and political rights, such as the right to life as in India or the notion of human dignity like in Germany. Thus, how can we find an approach to a human right to water that reflects its theoretical foundations, its practical implementation in the different legal orders and its normative content most appropriately?

As an orientation and guideline, the study employs the criteria developed by General Comment No. 15 to describe the content of the right to water (similar to criteria the Committee regularly applies in its General Comments): the human right to water entitles everyone to "sufficient, safe, acceptable, physically accessible and affordable water for personal and domestic uses."[71] I have argued before that General Comment No. 15 is not a legally binding document *per se*; however, the therein suggested structure and content of the right, if so accepted, would coherent and balanced. Thus, I see no harm in using the General Comment despite my previously expressed skepticism about its legal status as a tool to identify those elements of the right that would need to be covered, if the right was to be meaningfully accepted. The five attributes—sufficient, safe, acceptable, physically accessible and affordable—correspond with the three main characteristics of the content of the right to water, also developed in General Comment No. 15: availability, quality and accessibility.[72]

In order to understand the unique architecture of this right, one should start in its core, before examining its specific facets. In its very core, the human right to water, like every human right, derives from human dignity. The International Covenant on Civil and Political Rights (ICCPR), as well as the ICESCR, both emphasize this: The preambles of both the ICCPR and ICESCR state that "recognition of the inherent dignity and of the equal and inalienable rights of all members of the human family is the foundation of freedom, justice and peace in the world" and recognizes "that these rights derive from the inherent dignity of the human person".[73] Thus, the same must be true for a human right, even if it is not explicitly

[71] ESCR-Committee, *General Comment No. 15*, par. 2.

[72] *Ibid*., par. 12.

[73] ICESCR and ICCPR, Preamble, par. 2 and 3.

mentioned in the two covenants, but would be derived from covenant rights. This is not making a specific statement about the right to water itself: it is just recognizing that it exists for the same reason as all human rights, which is to respect the inherent dignity of the human person,[74] and that derived rights share the legal nature of their parent rights.

Leaving this general assumption about human rights behind, the question is what the specific nucleus of the right to water is. What is its essence, what is the raison d'être of the right to water? And vice versa, what is its greatest offense?

Ultimately, the answer can only be: to protect human survival. Its grossest offense, in turn, is human dying caused through the complete denial of access to water. Compared with this very essence, all other issues appear peripheral. The link between life and water could not be more crucial: human life simply cannot persist without water. Thus, any legal basis for a human right to water that does not take this most crucial link into account cannot be complete. I argue that this close link is also mirrored in the criteria outlined by General Comment No. 15. In the language of the General Comment, it is the criterion of "availability" that mainly corresponds with this nucleus of the right to water. Before any questions of quality and availability even arise, there must be access to sufficient drinking water, in a sufficient quantity for human life. The right to life must thus not be underestimated when discussing the legal sources of a human right to water.[75]

General Comment No. 15 also refers to a certain "quality" that water must hold in order to fulfill the right to water. In the committee's words, water must be "safe", thus "free from micro-organisms, chemical substances and radiological hazards that constitute a threat to a person's health" and "acceptable" in color, odor and taste.[76] The quality of the right to water speaks to the second parent right of the right to water: the right to the highest attainable standard of health. Already in General Comment No. 14,[77] the direct predecessor of the General Comment dedicated to water, the committee had identified many links between water and the right to the highest attainable standard of health. Access to potable water finds mention all over General Comment No. 14.[78] The fact that it is not mentioned in par. 3, which outlines the other human rights to which the right to health is related and upon which it is dependent, is simply due to the fact that the official recognition of the right to water was at that time yet to come.

The two adjectives that the Committee attributes to the criterion "quality" are safe and "acceptable". Only water that meets these two characteristics does not

[74] This holds true, of course, only if one accepts a sort of natural school of human rights (see discussion in Sect. 3.3.2 of this book), namely that human rights are not entirely man-made.

[75] McCaffrey (1992), pp. 7 ff.; Gleick (1998), pp. 487 ff.; Laskowski (2010), pp. 158 ff.

[76] ESCR-Committee, *General Comment No. 15*, par. 12(b).

[77] See ESCR-Committee, *General Comment No. 14: The right to the highest attainable standard of health (Art. 12)*, HRI/GEN/1/Rev.9 (Vol. I), 27 May 2008, at 78 (hereinafter "ESCR-Committee, *General Comment No. 14*").

[78] *Ibid*., par. 4, par. 11, par. 12(a), (b), (d), par.15, par. 34, par. 36, par. 40, par. 43(c), par. 51 and par. 65.

endanger human health. This element is similar, but at the same time, distinctly different, from the before discussed element of "availability". If water is not available at all, the human body runs the risk of fatal dehydration, thus the right to life is at stake. If water is generally available, but contaminated by micro-organisms or chemical substances, the risk for the human is predominantly a different one: it is not the risk of de-hydration, but rather that of infection or (non-lethal) toxication.[79] These risks mainly relate to the right to health rather than to the right to life directly (which is at stake only in most extreme cases). This "quality"-element of the right to water explains its second parent right: the right to the highest attainable standard of health.

Thirdly, the ESCR-Committee demands not only the "availability" and the "quality" of water, but also its "accessibility". Accessibility means physical and economic accessibility.[80] Physical accessibility implies that water must be within physical reach for each part of the population and in the immediate vicinity of each household; economic accessibility requires water to be affordable for each household.[81] This accessibility-element speaks to the last parent right of the right to water: the right to an adequate standard of living.

The right to an adequate standard of living, applied to the question of drinking water, demands that access to water must not compromise or threaten the realization of other covenant rights. On the one hand, this would be the case in terms of time and effort, if access to water was not within reasonable reach for a household, because fetching water may then become too time-consuming and exhaustive for the effective realization of other rights. The committee recognized this close link between housing (as explicitly mentioned part of the right to an adequate standard of living) and water already in an earlier General Comment,[82] later also confirmed by the United Nations (UN) Special Rapporteur on Adequate Housing.[83] On the other hand, access to water would impede the realization of other covenant rights, if paying the water bill consumes a too large part of the overall household income. The right is violated, if the costs of accessing drinking water constrains a household in a way that other rights—which are also costly to realize—are severely impeded

[79] In the case, a liquid is contaminated to an extent that it does not fulfill the human need for water in a way to prevent dehydration, this liquid must, in fact, be considered a "non-water" substance. If only such a "non-water" substance is available, it is, indeed, a question of (non-) availability rather than quality.

[80] The third element of accessibility as outlined by ESCR-Committee, *General Comment No. 15*, non-discrimination, is equally important, but does help identifying the legal sources of the right to water.

[81] ESCR-Committee, *General Comment No. 15*, par. 12 (c) i and ii.

[82] ESCR-Committee, *General Comment No. 4: The right to adequate housing (Art. 11 par. 1)*, HRI/GEN/1/Rev.9 (Vol. I), 27 May 2008, at 11, par. 8 (b) states: "An adequate house must contain certain facilities essential for health, security, comfort and nutrition. All beneficiaries of the right to adequate housing should have sustainable access to [. . .], safe drinking water, [. . .]."

[83] See, for instance, UN Commission on Human Rights, *Report of the Special Rapporteur on adequate housing as a component of the right to an adequate standard of living, Miloon Kothari*, E/CN.4/2004/48, 8 March 2004, par. 4.

or entirely hindered. Thus, those aspects of the right to water speaking to its physical and economic accessibility are best captured by understanding them as part of the right to adequate standard of living.

Altogether, we have identified three specific parent rights of the derived right to water: the right to life, constituting the right's core link to ensuring human survival, and a twofold mantle, one dealing with water quality in order to protect human health, the other dealing with adequate accessibility in order to ensure an overall adequate standard of living. One of the three categories of derivation, as identified before, is to form a synthesis of different rights, thereby creating a new derivative right. It appears that this is the most suitable option for understanding the right to water. It reflects best the normative content of the right to water, which also has three distinct elements: availability, quality and accessibility. The outcome is, in essence, a hybrid right: a right of different layers. In the case of the right to water, one of these layers derives from the right to life, the second is part of the right to an adequate standard of living, and a third one stems from the right to health.

This is an adequate solution for understanding the complex right to water, as the different claims under a "right to water" are very different in nature. Consumers in welfare States, whose household budgets are overly constrained by rapidly increasing water prices, appeal more to the socio-economic layer of the right. This is a substantially different concern from the one which humans in developing countries often face. If a person has no access to safe and sufficient water for nutrition at all, and has no means to receive this access, his status is not much different from a prisoner dying of thirst: the right to life is challenged. If all these different legal claims can be summarized under the umbrella of one right, this right can best be understood as a hybrid right. Often, the right to water has been considered either the one or the other: either socio-economic or civil–political. However, the truth lies somewhat in the middle: it is a bit of both.

There are three reasons why this hybrid nature of the right is often overlooked: firstly, an old understanding of the right to life; secondly, the wrong categorical assumption that a right must always be either civil–political or socio-economic; thirdly, a false assumption that a derivation from the right to life would create vast immediate positive obligations and a flood of legal claims to enforce them.

3.4.2.1 Antiquated Understanding of the Right to Life

The discussion on the technique of legal derivation has shown that a parent right and a derivative right must share certain characteristics. The derivative right inherits all its legal characteristics from its parent right. Undoubtedly, the right to water is a negative and a positive right at the same time—it not only sets out negative rules for States not to interfere with an existing water supply, it also imposes positive obligations on the State to protect and fulfill this right. The parental right for the right to water can therefore only be a right that itself is negative and positive in nature.

However, the character of the right to life as a positive right has long been contested. Traditionally, the prevailing view was that the right to life was a purely negative right. In this respect, it would not protect any person against death from famine, cold or lack of medical attention.[84] A State's toleration of malnutrition would therefore not constitute a violation of the right to life,[85] only the purposeful denial of food to persons entrusted in the State's care, such as prisoners, would be considered such a violation.[86] Some authors generally support a broader understanding of the right to life, while emphasizing that the right to life is just the precondition for the realization of other human rights, and that a right to subsistence should not be part of the right to life; this should purely belong to the category of economic, social and cultural rights.[87]

However, since the beginning of the 1990s, newer interpretations of the right to life go much further. Most prominently, the Human Rights Committee argues in favor of such a broader interpretation in its General Comment No. 6 on "The right to life".[88] According to General Comment No. 6, the right to life should not only protect against arbitrary deprivation of life, but also preserve the very minimum right of every human being to have the appropriate means of subsistence and a decent standard of life. The committee explicitly states that the protection of the right to life in this respect also requires States to "adopt positive measures".[89] This is only coherent: Firstly, arbitrary killings are only one of the mentioned aspects in Art. 6(1) ICCPR (namely its third sentence), whereas the overall scope of that article is much broader.[90] Secondly, it is not only in situations of State custody (such as imprisonment) in which people are incapable to realize their basic human rights themselves; the same applies in many developing countries. A distinction in this regard would be artificial.[91]

Thus, the conception of the right to life has, for good reasons, moved forward. This modern understanding of the right life makes it a suitable parent right for positive as well as for negative derivative rights. Thus, the nature of the right to life does not stand against the view that a very limited nucleus of the right to water can be derived from it.

---

[84] Przetacznik (1976), p. 587; Dinstein (1981), p. 114.

[85] Robinson (1958), p. 106.

[86] Dinstein (1981), p. 116.

[87] Cançado Trinidade (1991), p. 51.

[88] See CCPR-Committee, *General Comment No. 6: The right to life (Art. 6)*, HRI/GEN/1/Rev.9 (Vol. I), 27 May 2008, at 176; see in particular par. 1: "It is a right which should not be interpreted narrowly."; see support for this approach, for instance, at McCaffrey (1992), p. 9 who introduces the more narrow view with the word "unfortunately".

[89] *Ibid.*, par. 5.

[90] Bourquain (2008), p. 128; Winkler (2012), p. 51.

[91] Sepúlveda (2003), p. 147; Winkler (2012), p. 51.

3.4.2.2 Over-Categorization of Rights as Either Civil–Political or Socio-Economic

As pointed out in Chap. 2, one of the General Comment's (few) drawbacks is the fact that—due to its nature as an interpretation of the ICESCR—it focused on elaborating on the normative content of a human right to water; it relates to economic and social rights, while only mentioning briefly the right to life. In order to support the acceptance of the right in the future and in order not to undermine the General Comment's authority, it appears that many legal scholars jumped onto this bandwagon and dismissed any approach to the right to water different from the one expressed in General Comment No. 15. However, one thereby easily overlooks that there is no need to read the General Comment in a way suggesting that the right to water *only* derives from the right to an adequate standard of living (and the right to health). It was simply these rights that the General Comment, as an interpretation of the ICESCR, naturally focused on. However, assuming that these are the only elements of a right to water would be an incomplete approach to the normative content of the right.

This leads, in turn, to another point. There is a tendency in international law to strive for a clear distinction between civil–political and socio-economic rights. This is understandable in terms of order and legal clarity. However, it is not reflective of how interwoven most rights are with each other. These interconnections exist regardless of the classification of rights as civil–political and socio-economic rights. However, there is no apparent reason why a right in itself must always exclusively fall into one of the two categories; to the contrary, a right that is derived from several parent rights can easily be both, civil–political and socio-economic. The strict distinction that rights must be either in the one category exclusively or in the other is not always valid.

This is the extraordinary chance that lies in the fact that the right to water is mentioned in neither the ICESCR nor the ICCPR. This fact has often been bemoaned; however, one should at the same time embrace it. It allows for approaches that reflect the right with all its facets, attributing different legal claims to different parent rights i.e. to those that are most appropriate to a particular situation and that allow the appropriate legal remedy. This situation resembles in some sense that one of the Indian judges mentioned in Chap. 2[92]: had the Indian constitution stated that protection of water was a "directive principle", it would have been difficult to justify the recognition of a right to water as derived from the right to life. It was the silence of the law that gave the judiciary the chance to give the right the protection they thought was appropriate. Similarly, the unclearly defined legal status of the right to water in international law allows us to see it in a way which best captures its complexity: a right with several parent rights, some of them civil–political, some of them socio-economic.

[92] See discussion about the specific distinction in the Indian constitution between "fundamental "rights" and "directive principles".

3.4.2.3 False Fear of Creating Unforeseeable State Obligations

The reluctance to accept the right to life as one of the parental rights for the right to water might also be grounded in the fear of unforeseeable consequences. It has been argued that it appears paradoxical, if, under the ICCPR, States were obliged to make an immediate guarantee of the right to water, whereas the ICESCR only proscribes the duty to "progressively" realize the Covenant rights.[93] Such an immediate obligation would create extensive State liability, so the argument goes. However, this is nothing more than a common misunderstanding of the duties under the ICESCR as opposed to the duties under the ICCPR. Some duties under the ICESCR are also immediate. In particular the non-discrimination rule, the obligation to respect and the obligation to protect", as much as realizable through legislative means, are independent of the States' available resources, and thus of an immediate nature.[94] In addition, the ESCR-Committee employs the legal figure of "core obligations", as developed in the Limburg Principles[95] and General Comment No. 3,[96] to determine which other obligations are immediate. General Comment No. 15 has picked up this concept and identifies immediate core obligations for the case of the right to water.[97]

Where a person is deprived of water in a life-threatening manner, due to the State's interference (as an obligation to respect) or due to a third person's interference (as an obligation to protect), either the right to life is at stake or the "core obligations" under the rights to an adequate standard of living and the highest attainable standard of health. For the question of the State's concrete obligation it makes little difference: both obligations are immediate. In fact, the core obligations under the ICESCR go even further: they can capture cases in which a person is deprived of water in a non-life-threatening manner, but still trigger a core obligation of the State (e.g. cases of non-discrimination).[98] Thus, State's obligations would not be widened. For the most severe cases of water deprivation, it would just be a different right at stake: the right to life.

Accepting the right to life as one parent right of the right to water would therefore not create new obligations. It would just make existing obligations more enforceable, along two dimensions. Firstly, it would increase the circle of States to

[93] McCaffrey (2005), pp. 97–98.

[94] Klee (2000), p. 177; Rudolf (2007), p. 30.

[95] See UN Commission on Human Rights, *The Limburg Principles on the Implementation of the International Covenant on Economic, Social and Cultural Rights*, E/CN.4/1987/17, 8 January 1987, Annex.

[96] CCPR-Committee, *General Comment No. 3: Implementation at the national level (Art. 2)*, HRI/GEN/1/Rev.9 (Vol. I), 27 May 2008, at 174.

[97] See ESCR-Committee, *General Comment No. 15*, par. 39.

[98] In sum: Under a combined concept, as suggested in this book, cases in which water deprivation is life-threatening would create a violation of that part of the right to water which is derived from the right to life; cases in which water-deprivation is not life-threatening create a violation of that part of the right which is derived from the right to health; see also Rudolf (2007), p. 32.

be held liable for those most severe violations of a right to water (as there are more signature States to the ICCPR than to the ICESCR). Secondly, it would create stronger remedies for the individual. The weakness of individual enforcement mechanisms of socio-economic rights in international law is well known. To the extent that procedures exist at all, they appear more inquisitorial and supervisory in nature, relying on persuasion rather than coercion. Tying the core of the right to water to the right to life would open up the gates to the stronger remedies of the ICCPR. These are not only the periodic reports under Art. 16 ICESCR, but also those under Art. 40 ICCPR, which would have to include reporting on the water situation in a respective country. Individuals would also gain access to the communication procedure of the ICCPR for those cases in which a violation of the right of water jeopardizes their life. The absence of an equivalent procedure under the ICESCR has often been criticized, as it places significant limits on the committee's ability to develop jurisprudence and case-law and decreases the chance of victims to seek international redress for violations of socio-economic rights.[99] Recently, this call for an Optional Protocol to the ICESCR has been answered: After the hard work of the Open-Ended Working Group on an Optional Protocol to the International Covenant on Economic, Social and Cultural Rights,[100] the UNGA adopted Res. A/RES/63/117 in December 2008, introducing the long desired Optional Protocol that allows for individual communications with regard to violations of rights under ICESCR.[101] The mandatory ten ratifications[102] for the Optional Protocol to enter into force have very recently been reached.[103] However, the Optional Protocol of the ICCPR will, for a long while, remain the further-reaching remedy. The high number of ratifications of the Optional Protocol to the ICCPR, currently 114,[104] is most likely to remain out of reach for the Optional Protocol to the ICESCR for a long time to come.

3.4.2.4 Summary

For all these reasons, it appears appropriate to view the right to water not as a one-sided right, but rather as a right of different layers. It is not enforceable through only one certain court or under one specific procedure. It depends on the respective claim, as to which part of the right to water is at stake, and thus which procedures

[99] ESCR-Committee (1991).

[100] For an overview of their background, mandate and delivered reports, see webpage of the Office of the High Commissioner for Human Rights (OHCHR), http://www2.ohchr.org/english/issues/escr/intro.htm; for insightful background information, also see, Albuquerque (2010).

[101] UNGA, *Optional Protocol to the International Covenant on Economic, Social and Cultural Rights*, A/RES/63/117, 5 March 2009.

[102] *Ibid.*, see Art. 18.

[103] For the current status of number of ratifications, see the UN webpage at http://treaties.un.org/Pages/ViewDetails.aspx?src=TREATY&mtdsg_no=IV-3-a&chapter=4&lang=en.

[104] *Ibid.*

and courts are accessible for the individual. There is no rule and no reason that would prevent a right from being such a hybrid right. As long as the interest or the substance matter behind a right is reasonably coherent, there is good reason to address a variety of claims under the umbrella phrase of one right. One should only keep the complexity of the claims in mind that lies behind this common façade: some entitlements relating to life, some relating to an adequate standard of living, and others relating to health.

One question remains: has it been a mistake to label and promote the right to water over the last years on the international agenda as a socio-economic right exclusively? The answer, maybe surprising, is "no". Political and diplomatic reasons are strong to do so: as a purely civil and political right, there was the risk for States to only accept minimal claims under the right; and the way the human rights institutional system is currently laid out, rights must either be civil and political or socio-economic, so they do not fall through both categories. However, it is the task of a concise legal analysis like undertaken in this chapter to point out that this socio-economic labelling was politically and diplomatically motivated, rather than being conceptually required.

3.4.3 Contouring the Right: Water and Sanitation—One or Two Rights?

Recently, sanitation has more and more moved into the focus of global attention. The UN declared 2008 as "The International Year of Sanitation".[105] In 2009, Independent Expert, now Special Rapporteur, Catarina de Albuquerque chose the topic of sanitation as the "issue in focus" of the first year of her mandate.[106] Several non-governmental organizations (NGOs) have picked up this topic in the following and launched special programs on sanitation, some of which coordinated by the "Sustainable Sanitation Alliance".[107] Recent Millennium Development Goals (MDGs) Reports explicitly warn that MDG No. 7 with respect to sanitation will most likely be missed if a rapid acceleration of progress cannot be made within the next few years; the global challenges with regard to sanitation at present appear to be even steeper than those with regard to drinking water.[108] It is therefore an appropriate time to consider whether sanitation is part of the right to water, or whether it is an independent right.

---

[105] UNGA, *International Year of Sanitation, 2008*, A/RES/61/192, 6 February 2007.

[106] For the Expert's report see HRC, *Report of the independent expert on the issue of human rights obligations related to access to safe drinking water and sanitation, Catarina de Albuquerque*, A/HRC/12/24, 1 July 2009.

[107] For an overview of the coordinated programs, see http://www.susana.org/.

[108] See the alarming figures in UN (2009) with regard to sanitation, p. 45. See also UN (2012), pp. 52–55.

The strongest reason to assume that water and sanitation constitute one right is simply that they are so often dealt with together. Just to remember three prominent examples: MDG No. 7c calls to "reduce by half the proportion of people without sustainable access to safe drinking water and basic sanitation". Similarly, Catarina de Albuquerque received her mandate with the title "Independent Expert on the issue concerning human rights obligations related to access to safe drinking water and sanitation"; it was renewed as "Special Rapporteur on the human right to water and sanitation". The UNGA Resolution of 28 July, 2010, is entitled "The human right to water and sanitation", not the human right*s* to water and sanitation.

A second reason for combining the two topics in one right is that, whereas the right to water still gets much more attention than the right to sanitation, the challenges with regard to sanitation even exceed those with regard to drinking water. Thus, sanitation could benefit from the force and speed that the discussion on a right to water has developed, and stand, as a sort of legal "piggyback", higher chances of being more clearly recognized as a human right in the future.[109]

However, there are also reasons that stand against such a unified approach.[110]

Firstly, the listing of water and sanitation together in certain documents often has practical or political reasons; it is not an expression of the legal conviction that water and sanitation must be considered as one right. In the case of MDG No. 7, sanitation was left out in the original draft in 2000; only in 2002, on the occasion of the World Summit on Sustainable Development, were the targets with respect to sanitation added. As sanitation was a later addition to the original targets, it appeared more feasible and appropriate to include it in an existing goal, rather than creating an entirely new one. The scope of each Millennium Development Goal is generally rather broad: not everything that is summarized in one goal is automatically one right, given that for instance the fight against extreme poverty and the fight against global hunger are also summarized in MDG No. 1. Something similar can be said about the mandate of the Special Rapporteur. The fact that her mandate deals with obligations with respect to water and sanitation does not make this necessarily one right. There is no such mantra as "one expert, one right". Sometimes different aspects of one right are spread over several mandates, whereas in other cases one expert deals effectively with more than one right. Finally, with regard to important documents, one should note that General Comment No. 15, the by far most elaborate document on the normative content of the right, is explicitly named "The right to water", not "The Right to water and sanitation". The Comment mentions sanitation only a few times, as an aside to the main considerations.

Secondly, even the Special Rapporteur herself considers sanitation and drinking water as two distinct rights. In her report on sanitation, the (at the time) Independent

---

[109] Jenkins and Sugden (2006), p. 8.

[110] Also Winkler (2012) comes to this conclusion, pp. 127–129, albeit for quite different reasons from the ones I suggest here, namely that there are also non-water borne aspects of sanitation (p. 127) and that the technical requirements necessary for the implementation of both rights are different (pp. 128–129). These reasons are certainly not untrue. My account is just different in that I see the main reason to be the two rights' different origins, as will be explained in the following.

Expert found that "there is an ongoing discussion about sanitation as a distinct right" and that "recent developments in human rights law [...] suggest a trend towards recognition of such a distinct right", such that the expert "supports and encourages developments in line with this trend".[111] In this report, the expert also highlighted connections of the right to sanitation to other human rights. In this enumeration, the human right to water is named as one, as are life, dignity, health or education. This makes it clear that the expert sees close connections between water and sanitation, but considers them two distinct rights.

Thirdly, summarizing water and sanitation as one right bares the risk to create larger opposition to such an overarching right: those States that accept one element, but not the other, might decide to object to the establishment of the right as a whole. This is not an unlikely scenario. More and more States have moved towards accepting a human right to water, while still remaining skeptical about a right to sanitation. Just to give one example: Colombia acknowledged on the occasion of discussing the UNGA resolution of 2010 that water only has the character of a right when used for drinking, not if used for other purposes such as sanitation.[112]

Fourthly, and most importantly, both rights should be treated separately due to their distinctly different legal nature. Sanitation, more than many other human rights issues, evokes the concept of human dignity.[113] Denying access to sanitation infringes the very core of human dignity; therefore it is not sufficient to consider sanitation only as it relates to other human rights.[114] This emphasizes that sanitation in its core is most strongly linked to dignity, while the strongest core element for a right to drinking water is, as I have argued in this chapter, that of the right to life. In other words, the parent rights from which both rights derive are not entirely the same. Although, in their mantle, they are very similar—as both clearly relate to an adequate standard of living and to health—they are slightly different in their core. In the language of parent and derivative rights: without having the same parents, two rights cannot be seen as the same derivative right. They can only be legal half-siblings.

Thus, water and sanitation should conceptually not be regarded as one human right, but rather as two distinct human rights.

---

[111] See HRC, *Report of the independent expert on the issue of human rights obligations related to access to safe drinking water and sanitation, Catarina de Albuquerque*, A/HRC/12/24, 1 July 2009, par. 49.

[112] See for instance statement of the Colombian representative on the occasion of the adoption of Res. 64/292, UNGA, *General Assembly adopts resolution recognizing access to clean water, sanitation as human right*, Press Release GA/10967, 28 July 2010. Colombia eventually voted in favor of the resolution, but explicitly stated its understanding that only water for drinking was included.

[113] I have submitted a similar account of the relationship between water, sanitation and human dignity at Thielbörger (2013a), pp. 323–324.

[114] See HRC, *Report of the independent expert on the issue of human rights obligations related to access to safe drinking water and sanitation, Catarina de Albuquerque*, A/HRC/12/24, 1 July 2009.

3.4.4 *Different Legal Levels of Protection—Relationship of Constitutional Rights to Water and a Human Right to Water*

This chapter has pointed out by now a more differentiated approach to the right to water than often considered. However, one question remains: is there a need for such a right on the international level, particularly if we have already seen its protection in several national legal orders, as illustrated in Chap. 2 of this book? What would be the relationship between such a human right to water at the level of international law and the rights to water that already exist in national or regional legal orders?[115]

International law and human rights law can influence the reading and interpretation given to rights in national legal orders, both national and sub-constitutional law. This includes not only the classical discussion of the advantages of monism versus dualism.[116] It also speaks to the general tendency to interpret provisions of national constitutions in light of international law; for the German case, this approach has famously been called the "Völkerrechtsfreundlichkeit des Grundgesetzes"[117] (openness of the German constitution to international law), as applied by the German Constitutional Court, existing in similar form also in many other legal orders.

More specifically, the relationship between human rights and rights protected in national constitutions still occupies legal philosophers until today. Gianluigi Palombella notably pointed out the conceptual distinction between human rights and, as he calls them, fundamental rights[118]: the former being moral visions of what makes us human,[119] the latter being their institutional, ethical and political counterparts, carrying a meta-normative role in particular legal orders.[120] This distinction emphasizes the need for human rights, which in their broad vision often face difficulties in terms of concrete implementation, and therefore greatly benefit from their reflection in national fundamental rights.

These different rights in the national legal orders also reflect the different national legal and cultural traditions. As such, the question is not only what

---

[115] I have given a more detailed account of the relationship between the different national and international rights to water at Thielbörger (2013b), chapter 15. In addition to the two Kadi approaches, I therein also discuss a third possible approach to the relationship, namely that one suggested by the US Supreme Court, *Medellín v. Texas*, No. 06-984, 25 March 2008, 552 US 491 (2008)

[116] See for instance Kelsen (1958); see for a classical example of dualism Triepel (1923).

[117] Decision of the BVerfG, Case No. 1 BvR 636/68, 4 May 1971, BVerfGE 31, 58, 76.

[118] Palombella (2007). In terms of the different schools of human rights discussed above, it is clear that Palombella leans towards the natural school approaches.

[119] *Ibid.*, p. 402.

[120] *Ibid.*, pp. 409 ff.

human rights mean for the interpretation of rights in national legal orders. One must also consider, reciprocally, the meaning of national fundamental rights for international human rights. In this light, not only an interpretation of national law in the light of international law, but also, vice versa, a constitutional interpretation of international human rights law appears appropriate.[121]

The right to water is an excellent example to demonstrate the benefits and usefulness of such a reciprocal approach. If a human right is mainly connected to a rather abstract value, and not at the same time connected to the delivery of a certain good, different national systems share a more coherent idea, or in the words of Palombella, a more common "moral vision", about what the protection of the specific right entails. In the case of freedom of speech, for instance, the common understanding would be that a guarantee of a (relatively) free press, (relatively) free access to information and the possibility of unthreatened public demonstration are needed. Otherwise, a country is generally not considered to respect freedom of speech. However, the case of the right to water is even more complex. The question of how to shape adequate access to drinking water must be answered entirely differently in various cultural circles. The human right to water cannot, for instance, on its own answer the question of how much water needs to be accessible daily for each human being. This depends significantly on the climate, which varies from country to country, and on the general health conditions in a given country. Furthermore, access to water is interpreted differently in different cultures: whereas African and Asian rural cultural communities heavily rely on dwells and natural waterholes, access to water in western countries in order to be "adequate" entails technically sophisticated water supply technologies. In some cultural communities, access to water is exclusively understood as access to running water from a water-tap, in other communities, excluded from the infrastructure necessary to make ordinary running water "safe", the understanding entails drinking water out of bottles. The expectations of a right to water in water-rich countries like Finland or Canada necessarily differ from the expectations of populations in sub-Saharan Africa.[122]

These and other cultural differences cannot sufficiently be addressed in a universal human right alone. They find reflection also in the protection of the right to water in national legal orders. It is ultimately the task of the national legislator and of national courts to shape the right, according to the needs and conditions of their national communities. However, this plea for a constitutional reading of human rights does not undermine the concept of human rights. To the contrary, it strengthens it, making it more credible in diverse national orders. Strengthening the meaning of constitutional rights does not make human rights

[121] A similar approach for the field of International Economic Law is suggested by Petersmann (2008).

[122] Despite being different in their needs, communities have the great potential to learn from each other. See in this respect the recent collection of good practices published by the Special Rapporteur's Office: Albuquerque and Roaf (2012).

obsolete. It is again MacCormick's dilemma[123] that gives us the reason: if there was no supra-national approach to the essential human need for water, universally applicable, the right would most likely not be protected precisely where it is needed most: those to whom the right is most crucial, would be those who lack it. To prevent this is exactly one of the most important tasks of a right to water in international law.

3.4.4.1 The Relationship of Different Legal Orders After the Kadi-Case

In shaping these different standards of protection, different legal orders can, however, begin to conflict. This problem has most famously been addressed in the recent cases of *Kadi*[124] where the European Court of Justice (ECJ) (now: Court of the European Union) ultimately was concerned with a clash of international and European law. Similar questions can arise for the case of international and national law. The relationship of regional and national legal orders within the international legal order, and their potential clash, cannot be convincingly determined without taking this most important judgment into account. It has largely re-designed the view on the relationship of different legal orders.

Concerning the basic background of the case,[125] Mr. Kadi—a Saudi national—had been designated by the Sanctions Committee, an organ of the UN Security Council (UNSC), as being associated with Al-Qaeda. UN Member States were obliged to freeze the funds and other financial resources controlled by such persons. In order to give effect to those resolutions within the European Community (EC), the Council adopted a regulation[126] ordering the freezing of the funds of all persons named in a list annexed to that regulation. The list adopted by the Council was identical to the one created by the Sanctions Committee (and thus included Mr. Kadi).

Mr. Kadi claimed before the Court of First Instance (CFI) (now called: General Court), *inter alia*, the regulation infringed his right to property and his rights of defense, as protected under general principles of EC law. He had not been informed about the listing beforehand. In its judgment of September 2005, the court rejected

[123] See discussion on MacCormick (1976).

[124] ECJ, Joined Cases C-402/05 P and C-415/05 P, *Yassin Abdullah Kadi and Al Barakaat International Foundation v. Council of the European Union and Commission of the European Communities*, Judgment of 3 September 2008, [2008] ECR I-6351.

[125] I have given a more extended description of the factual background and analysis of the case at Thielbörger (2013b), chapter 15.

[126] Council of the European Union, *Council Regulation (EC) No 881/2002 of 27 May 2002 imposing certain specific restrictive measures directed against certain persons and entities associated with Usama bin Laden, the Al-Qaeda network and the Taliban, and repealing Council Regulation (EC) No 467/2001 prohibiting the export of certain goods and services to Afghanistan, strengthening the flight ban and extending the freeze of funds and other financial resources in respect of the Taliban in Afghanistan*, No. 881/2002, 27 May 2002, OJ 2000 L 139, 9.

all of Mr. Kadi's pleas, and confirmed the validity of the regulation. The court argued that the European courts had no jurisdiction to review the validity of the regulation in the first place, except for *ius cogens* violations.[127] Member States were obliged to comply with the resolutions of the UNSC according to the UN Charter, which, as an international treaty, ultimately must prevail over Community law, as followed from Art. 103 UN Charter. The court thereby accepted in essence the "primacy" of international over European law. Its reasoning, in particular, was based on the old Art. 307 EC, which stated that the provisions of EC law should not affect the validity of international obligations incumbent upon the Member States prior to the EC's creation (including of course, the UN Charter dating from 1945).

In deciding on Mr. Kadi's appeal, the ECJ (now called Court of Justice of the European Union) found that, contrary to the findings of the CFI, the European courts *did* have jurisdiction to review the internal lawfulness of the regulation contested by Mr. Kadi. It even regarded the review of the regulation with regard to its conformity with fundamental rights as a constitutional guarantee from the EC Treaty, which constituted an autonomous legal system that could not be prejudiced by any international agreement. The ECJ concluded that the Community courts must ensure the full legal review of all Community acts in the light of fundamental rights, as they formed an integral part of the general principles of European law.[128] This was necessary also for Community measures which, like the contested regulation, were only designed to give effect to resolutions adopted by the UNSC. At the same time, the court insisted that, in reviewing the contested regulation, it was not reviewing the validity of the UNSC resolution itself (admittedly a rather difficult claim to sustain given that the outcome of the review could be to render the resolution inoperable in its European Union (EU) context).[129]

With this approach established, the Court of Justice the EU found Mr. Kadi's rights violated.[130] Mr. Kadi had not been notified of the grounds on which the inclusion of his name was based at the time of the inclusion or immediately after.

---

[127] As stated by the CFI, the Court had "jurisdiction to review the lawfulness of the contested regulation and, indirectly, the lawfulness of the resolutions of the Security Council at issue, in light of higher rules of international law falling within the ambit of *ius cogens*, in particular the mandatory prescriptions concerning the universal protection of the rights of the human person." CFI, Case T-315/01, *Yassin Abdullah Kadi v. Council of the European Union and Commission of the European Communities*, Judgment of the Court of First Instance of 21 September 2005, [2005] ECR II-3649, par. 282.

[128] The statement of the Advocate-General in the case, Miguel Poires Maduro, is informative in this respect. He argues that "the Court cannot, in deference to the views of those institutions, turn its back on the fundamental values that lie at the basis of the Community legal order and which it has the duty to protect. Respect for other institutions is meaningful only if it can be built on a shared understanding of these values and on a mutual commitment to respect them." ECJ, Joined Cases C-402/05 P and C-415/05 P, *Yassin Abdullah Kadi and Al Barakaat International Foundation v. Council of the European Union and Commission of the European Communities*, Opinion of Advocate General Maduro of 23 January 2008, [2008] ECR I-6351, par. 44.

[129] *Ibid.*, par. 287.

[130] *Ibid.*, par. 328.

This would have been necessary in order to enable him to exercise his defensive rights. The court also found that the freezing of funds, in the way they were carried out, constituted an unjustified restriction of Mr. Kadi's right to property. The court annulled the Council regulation in so far as it applied to Mr. Kadi (although its effects were maintained for a period of 3 months in order to allow the Council to remedy the infringements found). In essence, the ECJ thereby accepted the general primacy of international law over *secondary* Community law, but refused to expand this primacy of international law to *primary* Community law (in particular to its general principles such as the protection of fundamental rights).

There are, of course, good arguments for the positions adopted by both courts.[131] One is to accept the primacy of international law, in particular the UN Charter, in all circumstances over national (or, as in the case of *Kadi*, regional) law, thereby strengthening the unity of the international legal order. This is an approach exemplified by the CFI, who argued that—if an EC measure implementing a UNSC resolution was to be reviewed—it could be done only on the basis of international legal standards themselves. The opposite approach is to take a more nuanced view, distinguishing between primary and secondary law, and allowing regional orders to provide more extensive protection, even if that might have the effect of bringing to question the primacy of the original international norm. Applying the *Kadi*-case, and the two different approaches taken by the two European courts onto the case of the right to water protected on different legal levels, leads to two possible scenarios based on the two different approaches.

3.4.4.2 National Level of Protection Exceeding International Level of Protection

It is possible that the protection offered by a domestic legal system exceeds the level of protection under international law (similar to the scenario in Kadi). Consider for instance a case in which national law forbids the privatization of water supply systems and where the national legal conviction is that water should be supplied through the State only.[132] Imagine a regional community in such a country decides, nevertheless, to privatize its regional water supply system, in order to meet standards of water quality as proscribed by international law. These

---

[131] *Kadi* has, of course, produced a fierce academic debate (siding either with the CFI or the ECJ), particularly disagreeing over the judgments' conception of the relationship between international and national law. For some of the most influential articles, see Eckes (2008); de Búrca (2009); Isiksel (2010); Halberstam and Stein (2009); Scheinin (2009); Tomuschat (2009); Ziegler (2009); Besson (2009).

[132] This example is, on purpose, held in the abstract, in order to emphasize the generality of this scenario. That such convictions generally exist, is, for instance, shown by the examples of the Netherlands and Uruguay which have both banned the privatization of water supply systems by law; see for these examples Hall et al. (2004); for the example of the Netherlands, see also Laskowski (2010), pp. 681–686.

standards say that the international rules must be met under all circumstances, be it by reaching out for private money. Could citizens of that community claim at a national court that their right to water, as understood in the national legal tradition, is violated through the privatization by relying in international norms? Or would ultimately the legal conviction in international law prevail, which regards privatization of water as generally compatible with the human right to water?[133]

In line with the views of the CFI, one would accept the general primacy of the UN Charter as well as the primacy of international *ius cogens*. One would argue that, although water privatization was banned in domestic law, the community aimed to reach international water quality standards and did so by water privatization as a means generally accepted in international law. Annulling the privatization would mean annulling a decision between two competing values that has already been taken in the realm of international law itself (according to which neither public nor private delivery is preferable). However, the human right to water is, as suggested in this chapter, derived from rights contained in the ICCPR and the ICESCR, not from the Charter itself. Only the very essential part of the human right to water which derives from the right to life can be considered part of ius cogens (certainly the ICESCR is not considered part of it).[134] Only this element of the right would prevail over the domestic ban on privatization. Here it shows that identifying the parent right of the right to water is crucial: only if (partly) derived from the right to life (which can be considered *ius cogens*), can the right to water prevail over diverging national law, if one follows the CFI's reasoning.

If adopting the view of the ECJ in Kadi, one would need to distinguish between primary and secondary law in order to determine which legal order prevails. Assuming that a right to water is part of the canon of national primary law or fundamental rights (whether this is the case, varies from country to country, as we have seen in Chap. 2), the courts would find a violation of national primary law, which could not be justified by the attempt to implement international law standards. Otherwise though, international law would trump deviating ordinary domestic law: the privatization would be found lawful by the courts, as it was explicitly allowed under international law. In this approach, the way in which the right is domestically recognized becomes crucial: only if the right is part of domestic primary law, can it trump diverging international standards.

3.4.4.3 International Level of Protection Exceeding National Level Protection

The opposite case could be one in which the protection of international law goes further than the one accepted in national law. Consider for instance a case in which the human right to water generally views it as illegal to cut down an existing means

[133] Albuquerque and Winkler (2010), pp. 167 ff.

[134] Laskowski (2010) comes to a similar conclusion, p. 896.

of water access and assume further that a certain minimum amount of water (e.g. 20–25 L) must be ensured under all circumstances. Consider at the same time, however, that a national State has, according to its own domestically accepted right to water, legalized a practice to cut down water supplies, if water bills were not paid over a certain period of time and the household has been notified; or that the national legislator considers a lower amount of water (e.g. 10 L) than the one suggested by international law as sufficient. Would citizens be successful with their legal claims, if they challenged this State practice in court which does not meet international standards?

Following the line of reasoning of the CFI in *Kadi*, the human right to water would prevail over conflicting domestic law, again, only if it can be seen as part of the UN Charter or as part of *ius cogens*. Given its origins in the ICESCR and ICCPR, only a small nucleus of the right (that one necessary for bare survival) would prevail over the domestic provisions.

However, the ECJ's approach would suggest that the restrictive understanding of the right to water in the national legal order would prevail, if it was part of domestic primary law (e.g. a recognition of the right in the constitution). If, and only if, the State's interpretation of the right was justified under national constitutional law, the restrictive practices would be upheld, even if they deviated in the negative from the right to water on the international level.

3.4.4.4 Avoiding the Dilemma: Human Rights as Minimum, Not Uniform Standards

So which of the two approaches is preferable and which maximizes the protection of the right to water?[135]

On the one hand, the CFI suggests an absolute supremacy of international law, seemingly strengthening the unity of the legal orders (with the UN Charter at the top). On the other hand, such an approach puts fundamental domestic law too easily aside. Like in *Kadi*, this can have the effect that a higher level of protection is easily put aside. The CFI does not distinguish within domestic law: laws of a higher nature are equally subordinated to international law as secondary laws—regardless of their different levels of (democratic) legitimization. An approach along these lines overlooks that the vivid interplay of the different legal levels holds the potential to optimize the protection of human rights. International law is always highly dependent on the actors in, and the functionality of, domestic law, given the latter provides far more efficient legal enforcement mechanisms. An approach that reconciles international law with domestic law is needed. An approach such as the ECJ's in *Kadi* points exactly in this direction. It strikes an appropriate balance by distinguishing between primary and secondary domestic law to decide which law enjoys supremacy. The integrity of the domestic legal order is preserved, while

[135] I have submitted an extended account of this question in Thielbörger (2013b), chapter 15.

the general supremacy of international law is not generally challenged. Such a differentiation is the only logical consequence of the legislator's decision to give the right a constitutional status—or not to do so.

Two logical outcomes of this approach deserve further scrutiny. Firstly, where a national legal order deviates in the negative from the human right to water, the courts will apply this domestic constitutional provision offering lower levels of protection. From a human rights perspective, this is a dissatisfying outcome. However, one should not forget that in such a case the legislator finds itself in breach of international law (and can address this violation by either amending its constitution or its international obligations). Which of these two options to choose is, however, for the national legislators to decide, not for the courts. Where domestic primary law, however, is silent on the right to water, international law standards must be applied by domestic courts, even where an ordinary law explicitly tries to set out a divergent or lower standard with regard to water.

Secondly, cases in which domestic ordinary law exceeds the international human right to water might hold a paradoxical result. Bluntly applying the ECJ's approach would mean to prioritize the higher level of protection only if it is constitutionally guaranteed, not if such standards derive from ordinary law alone. Here, a difference between the *Kadi*-case and the water examples I have depicted in this chapter becomes apparent: while in *Kadi*, different rights clashed, in the examples I have depicted in this chapter different recognitions of the same right were at stake (e.g. different quality or quantity requirements or different modes of service provision). In solving these cases, it can make no difference whether a protection going beyond the human right to water is rooted in constitutional or ordinary law. Any other approach would contradict the purpose of international human rights. The ECJ's reasoning in *Kadi* applied to human rights such as the right to water is thus best understood as establishing that domestic measures cannot *undergo* a certain level of fundamental rights protection. They can, however, exceed this level. The human right to water thus establishes a minimum, not a uniform standard for the protection of the right to water.

3.5 Conclusion

This chapter has addressed the major theoretical questions about a human right to water: Can *water* be the content of a right? Is the creation of a *right* the appropriate approach to the human need for water? Would such a right correctly be described as a *human* right? Although some major objections behind these questions are legitimate, we have not identified one single of the objections as irrefutable. To the contrary, this chapter has highlighted the major benefits that a human right to water would entail, and elaborated an approach that reflects carefully the experiences in the different legal orders, avoiding the possible objections pointed out before.

The right to water is a right of highest complexity. That is why I have deemed it best to reflect it as a right of several layers, one of which is closely linked to the right to life (an often neglected element of the right), with the others layers being part of the right to an adequate standard of living and the right to the highest attainable standard of health. In this way, the sometimes bemoaned insufficient clarity in international law about the legal character of the right is not only a curse, but at the same time a chance: a chance to understand the right in a more nuanced way that is consistent with existing approaches in different legal regimes. This composed nature of the right should also lead to a more nuanced legal treatment of claims, currently gathered under the roof of one "right to water": some of these claims are effectively enforceable under the ICCPR; others have to be treated as socio-economic claims and will have to wait until the Optional Protocol to the ICESCR enters into force between more States than currently the case.

The other key question regards the relationship between such a human right to water on the international level and the rights to water accepted in the different national legal orders. In interpreting the case of *Kadi* and applying the ECJ's approach (rather than the CFI's one) to the case of water, I have drawn two main conclusions in this chapter. Firstly, where a domestic primary law recognition of a right to water diverges from international standards, these lower or diverging standards nevertheless prevail. We might not like this outcome from a human rights perspective, but it is inevitable in order not to disregard fundamental decisions of domestic constitutional legislators. Those legislators are, of course, urgently called upon to bring their constitutional law and their international human rights obligations in accordance (and will in most cases have a strong interest to do so). Secondly, the human right to water sets basic minimum standards which States cannot go beneath; they can, however, make more ambitious commitments to a right to water, regardless of whether they do so in their constitutional or ordinary law. This approach gives room for a constitutional interpretation of the human right to water—an interpretation through the lens of the existing national and regional legal orders.

Having established this refined understanding of a right to water, it would be more accurate to talk about "rights" to water, as suggested in the title of this book, than about one right to water alone. Firstly, already the supposedly one right to water in international law is a highly complex right, combining various legal claims of a very different nature under its roof (some civil–political, some socio-economic in nature). Secondly, the rights accepted in national legal orders sometimes go beyond what is accepted on the level of international law, thereby contributing to the shape of the overall right. In reading all these mosaic pieces together, one can indeed talk about a human right to water. However, this language should not come without the realization that, in fact, this one right is a conglomerate of many different rights in many different legal orders.

References

Albuquerque C (2010) Chronicle of an announced birth: the coming into life of the optional protocol to the international covenant on economic, social and cultural rights – the missing piece of the international bill of human rights. Hum Rights Q 32(1):144–178

Albuquerque C, Roaf V (2012) On the right track: good practices in realising the rights to water and sanitation. Office of the High Commissioner for Human Rights. http://www.ohchr.org/Documents/Issues/Water/BookonGoodPractices_en.pdf

Albuquerque C, Winkler IT (2010) Neither friend nor foe: why the commercialisation of water and sanitation services is not the main issue for the realization of human rights. Brown J World Aff 17(1):167–179
Alston P (1984) Conjuring up new human rights: a proposal for quality control. Am J Int Law 78 (3):607–635
An-Na'im A (2004) To affirm the full human rights standing of economic, social and cultural rights. In: Ghai Y, Cottrell J (eds) Economic, social and cultural rights in practice – the role of judges in implementing economic, social and cultural rights. Interrights, London, pp 7–16
Arnold C (1978) Analyses of rights. In: Kamenka E, Tay AES (eds) Human rights. E. Arnold, Port Melbournce, pp 74–86
Baxi U (2006) The future of human rights. Oxford University Press, New York
Bentham J (1843) Anarchical fallacies. In: Bauring J (ed) The works of Jeremy Bentham, vol II. W. Tait, Edinburgh
Besson S (2009) European legal pluralism after Kadi. Eur Constit Law Rev 5(2):237–264
Bleisch B (2006) The human right to water – normative foundations and ethical implications. Ethics Econ 4(2). http://www.ethique-economique.net
Bourquain K (2008) Freshwater access from a human rights perspective: a challenge to international water law and human rights law. Martinus Nijhoff, Leiden
Broughton J (2002) Descartes's method of doubt. Princeton University Press, Princeton
Brown W (2004) The most we can hope for...: human rights and the politics of fatalism. South Atl Q 103(2/3):451–463
Búrca G (2009) The EU, the European court of justice and the international legal order after Kadi. Harv Int Law J 51(1):1–49
Cahill A (2005) The human right to water – a right of unique status': the legal status and normative content of the right to water. Int J Hum Rights 9(3):389–410
Campbell T (2006) Rights: a critical introduction. Routledge, Abingdon
Cançado Trinidade AA (1991) The parallel evolutions of international human rights protection and of environmental protection and the absence of restrictions upon the exercise of recognized human rights. Rev Inst Interamericano Derechos Humanos 13:35–76
Coomans F (2002) The role of the UN committee on economic, social and cultural rights in strengthening implementation and supervision of the international covenant on economic, social and cultural rights. Verfassung Recht Übersee 35(2):182–200
Cranston M (1967) Human rights – real and supposed. In: Raphael DD (ed) Political theory and the rights of men. Indiana University Press, Bloomington, pp 43–53
Curley EM (1999) Descartes against the skeptics. Harvard University Press, Cambridge
Dembour MB (2006) Who believes in human rights? Reflections on the European convention. Cambridge University Press, Cambridge
Dembour MB (2010) What are human rights? Four schools of thought. Hum Rights Q 32(1):1–20
Descartes R (1999) Discourse on method and meditations on first philosophy, 4th edn (trans: Cress DA). Hackett Publishing Company, Indianapolis
Dinstein Y (1981) The right to life, physical integrity and liberty. In: Henkin L (ed) The international bill of rights. Columbia University Press, New York, pp 114–120
Donnelly J (2003) Universal human rights in theory and practice, 2nd edn. Cornell University Press, Ithaca
Dworkin R (1984) Rights as trumps. In: Waldron J (ed) Theories of rights. Oxford University Press, Oxford, pp 153–167
Eckes C (2008) Judicial review of European anti-terrorism measures – the *Yusuf* and *Kadi* judgments of the court of first instance. Eur Law J 14(1):74–92
Committee on Economic, Social and Cultural Rights (ESCR-Committee) (1991) Fact Sheet No. 16 (Rev. 1). http://www.ohchr.org/Documents/Publications/FactSheet16rev.1en.pdf
Franklin B (1793) Poor Richard's Almanack. Quoted in: McNeill JR (2001) Something new under the sun: an environmental history of the twentieth-century world. W W Norton, New York
Friauf KH (1971) Zur Rolle der Grundrechte im Interventions- und Leistungsstaat. Deutsches Verwaltungsblatt 86(18):674–682
Gleick PH (1998) The human right to water. Water Policy 1(5):487–503

Graham P (1996) The will theory of rights: a defense. Law Philos 15(3):257–270
Habermas J (1996) Between facts and norms: contributions to a discourse theory of law and democracy. MIT Press, Cambridge/MA
Halberstam D, Stein E (2009) The United Nations, the European Union and the king of Sweden: economic sanctions and individual rights in a plural world order. Common Mark Law Rev 46 (1):13–72
Hall D, Lobina E, de la Motte R (2004) Making water privatisation illegal: new laws in Netherlands and Uruguay. http://gala.gre.ac.uk/3769/1/PSIRU_9343_-_2004-11-W-crim.pdf
Hart HLA (1954) Definition and theory in jurisprudence. Law Q Rev 70(1):37–60
Hart HLA (1973) Bentham on legal rights. In: Simpsons AWB (ed) Oxford essays in jurisprudence. Second series. Clarendon, Oxford, pp 171–201
International Institute for Democracy and Electoral Assistance (IDEA) (2012) Compulsory voting – what is compulsory voting? http://www.idea.int/vt/compulsory_voting.cfm
Ignatieff M (2001) Human rights as politics and idolatry. Pinceton University Press, Princeton
Isiksel NT (2010) Fundamental rights in the EU after Kadi and Al Barakaat. Eur Law J 16 (5):551–577
Jenkins MW, Sugden S (2006) Rethinking sanitation: lessons and innovation for sustainability and success in the new millennium. UNDP Human Development Report Office, New York
Kelsen H (1958) Die Einheit von Völkerrecht und staatlichem Recht. Zeitschrift ausländisches öffentliches Recht Völkerrecht 19:234–248
Klee K (2000) Die progressive Verwirklichung wirtschaftlicher und sozialer und kultureller Rechte. Eine Interpretation von Art. 2 Abs. 1 des Internationalen Pakts für wirtschaftliche, soziale und kulturelle Rechte. Boorberg, Stuttgart
Ladwig B (2007) Kann es ein Menschenrecht auf Wasser geben? In: Rudolf B (ed) Menschenrecht Wasser? Peter Lang, Frankfurt am Main, pp 45–58
Laskowski SR (2010) Das Menschenrecht auf Wasser: Die rechtlichen Vorgaben zur Sicherung der Grundversorgung mit Wasser und Sanitärleistungen im Rahmen einer ökologisch-nachhaltigen Wasserwirtschaftsordnung. Mohr Siebeck, Tübingen
MacCormick N (1976) Children's rights: a test-case for theories of rights. Arch Rechts Sozialphilosophie 62(3):306–316
MacFarlane LJ (1985) The theory and practice of human rights. Dartmouth Publishing Co. Ltd, Aldershot
McCaffrey SC (1992) A human right to water: domestic and international implications. Georget Int Environ Law Rev 5(1):1–24
McCaffrey SC (2005) The human right to water. In: Bernasconi-Osterwalder N, Brown Weiss E, de Boisson Chazournes L (eds) Fresh water and international economic law. Oxford University Press, Oxford, pp 93–115
McIntyre A (2007) After virtue: a study in moral theory, 3rd edn. University of Notre Dame Press, Notre Dame
O'Neill O (1996) Towards justice and virtue: a constructive account of practical reasoning. Cambridge University Press, Cambridge
Palombella G (2007) From human rights to fundamental rights. Consequences and conceptual distinction. Arch Rechts Sozialphilosophie 93(3):396–426
Perry MJ (1998) The idea of human rights: four inquiries. Oxford University Press, New York
Petersmann EU (2008) De-fragmentation of international economic law through constitutional interpretation and adjudication with due respect for reasonable disagreement. Loyola Univ Chicago Int Law Rev 6(1):209–248
Przetacznik F (1976) The right to life as a basic human right. Rev Droits l'Homme 9(4):585–608
Raz J (1984) On the nature of rights. Mind 93(370):194–214
Robinson N (1958) The universal declaration of human rights: its origin, significance, application, and interpretation. Institute of Jewish Affairs, World Jewish Congress, New York
Rudolf B (2007) Menschenrecht Wasser: Herleitung, Inhalt, Bedeutung, Probleme. In: Rudolf B (ed) Menschenrecht Wasser? Peter Lang, Berlin, pp 15–43

Scheinin M (2009) Is the ECJ ruling in Kadi incompatible with international law? In: Eeckhout P, Tridimas T (eds) Yearbook of European law, vol 28, pp 637–653

Sen A (2004) Elements of a theory of human rights. Philos Public Aff 32(4):315–356

Sepúlveda M (2003) The nature of the obligations under the international covenant on economic, social and cultural rights. Intersentia, Antwerp

Shestack JJ (1998) The philosophic foundations of human rights. Hum Rights Q 20(2):201–234

Shue H (1980) Basic rights: subsistence, affluence, and U.S. Foreign Policy. Princeton University Press, Princeton

Speed S (2008) Rights in rebellion: indigenous struggle and human rights in Chiapas. Stanford University Press, Stanford

Stammers N (2009) Human rights and social movements. Pluto, London

Thielbörger P (2013a) Wasser und Würde. In: Kapust A, Gröschner R, Lembcke OW (eds) Wörterbuch der Würde. UTB Fink, Paderborn, pp 323–324

Thielbörger P (2013b) The right to water and its effective multi-level protection. In: Cremona M, Hilpold P, Lavranos N et al (eds) Reflections on the constitutionalization of international economic law – Liber amicorum Ernst-Ulrich Petersmann. Martinus Nijhoff, Leiden, pp 553–569

Tomuschat C (2009) What relationship is there between the universal legal order under the auspices of the United Nations and the EU legal order? In: Eeckhout P, Tridimas T (eds) Yearbook of European law, vol 28, pp 654–663

Triepel H (1923) Les rapports entre le droit interne et le droit international. Recueil des Cours I:73–121

Tully S (2005) A human right to access water? A critique of general comment no. 15. Neth Q Hum Rights 23(1):35–63

Tully S (2006) Flighty purposes and deeds: a rejoinder to Malcolm Langford. Neth Q Hum Rights 24(3):461–472

United Nations (UN) (2009) The millennium development goals report 2009. United Nations, New York. http://www.un.org/millenniumgoals/pdf/MDG_Report_2009_ENG.pdf

United Nations (UN) (2012) The millennium development goals report 2012. United Nations, New York. http://www.un.org/millenniumgoals/pdf/MDG%20Report%202012.pdf

United Nations Development Programme (UNDP) (2006) Human development report 2006: beyond scarcity – power, poverty and the global water crisis. http://hdr.undp.org/en/media/HDR06-complete.pdf

United Nations Educational, Scientific and Cultural Organization (UNESCO) (2012) The 4th edition of the UN World Water Development Report (WWDR4). http://www.unesco.org/new/en/natural-sciences/environment/water/wwap/wwdr/wwdr4-2012

van Hoof G (1984) The legal nature of economic, social and cultural rights: a rebuttal of some traditional views. In: Alston P, Tomasevski K (eds) The right to food. Martinus Nijhoff, Utrecht, pp 97–110

Veigha de Cunha L (2009) Water: a human right or an economic resource? In: Lamas MR, Martinez-Cortina L, Mukherji A (eds) Water ethics: Marcelino botin water forum 2007. CRC Press, London, pp 97–113

Vierdag EW (1978) The legal nature of the rights granted by the international covenant on economic, social and cultural rights. Neth Yearb Int Law 9:69–105

Wasserstrom R (1964) Rights, human rights and racial discrimination. J Philos 61(20):628–641

Wellman C (1995) Real rights. Oxford University Press, Oxford

World Health Organization (WHO), United Nations Children's Fund (UNICEF) (2006) Meeting the MDG drinking water and sanitation target: the urban and rural challenge of the decade. http://www.who.int/water_sanitation_health/monitoring/jmpfinal.pdf

Winkler IT (2012) The human right to water: significance, legal status and implications for water allocation. Hart, Oxford

Ziegler KS (2009) Strengthening the rule of law, but fragmenting international law: the *Kadi* decision of the ECJ from the perspective of human rights. Hum Rights Law Rev 9(2):288–305

Chapter 4
Implementation: Independent Monitoring, Enforcement Against the Private Sector, and International Realization

This book has by now examined in Chap. 2 the current status of the right(s) to water in the different legal orders and identified its still not clearly defined status, in particularly in international treaty and customary law. Chapter 3 has raised conceptual objections against the idea of such a human right, replying and ultimately overruling all of these objections. It has also identified the most coherent and effective approach to the right to water, as a right of distinctly different layers and as a compound of national and international recognitions, which I have called the right(s) to water.

However, even if interpreted and construed in this way, we cannot simply assume that the right(s) to water would make a significant difference in practice. As Henry Shue phrased it, "just a proclamation of a right is not the fulfillment of a right, and may in fact be either a step toward or away from actually fulfilling the right".[1] Some empirical research indicates that other factors, amongst them structures of "good water governance", might be of significant importance in solving the global water problem,[2] potentially of much bigger importance than the recognition of the right itself. Thus, what conditions would enable the right to water to make the difference it strives to make, and what parameters would make it effective on the international level? Not asking these questions would mean to lead a discussion in an ivory tower; the findings of Chaps. 2 and 3 must rather be applied to practical fields.

This chapter suggests that three of the most crucial elements to make the right effective are independent monitoring, enforcement towards the private sector, and collective realization through international means. I understand these conditions not necessarily to be sufficient for the right to be effective; but without them, the right is certainly doomed to fail in practice.

---

*All cited webpages were last accessed on 1 August 2013.

[1] Shue (1980), p. 16.

[2] Grimes (2009); Pahl-Wostl et al. (2008); Solanes and Jouravlev (2006). The mere discussion on a right to water can, of course, stipulate these other factors.

DOI 10.1007/978-3-642-33908-0_4,

4.1 Independent Monitoring: The Case of the UN Special Rapporteur on the Human Right to Water and Sanitation

Turning to the first of these three elements, violations of a right to water must be identified in the first place. With the content defined and the obligations outlined in General Comment No. 15 (despite its ambiguous legal character a good guide to what the normative content of the right would be), the macro-level of this identification is more or less met: what generally constitutes compliance and what constitutes negligence of the human right to water is relatively clearly set out. However, concrete instances in which a State, or a corporation, violates the right at the micro-level must also be observed and reported. Thus, in order to make the right effective in practical terms, compliance and non-compliance of the right must be independently monitored.

The United Nations (UN) has understood the great importance of effective monitoring of human rights violations with respect to water. It created a new mandate within its organizational framework: Catarina de Albuquerque assumed the position as Independent Expert on the issue of human rights obligations related to access to safe drinking water and sanitation in 2008,[3] re appointed as Special Rapporteur on the human right to water and sanitation in 2011.[4]

4.1.1 The Monitoring Role of Experts

"Independent Experts", "Special Rapporteurs" or "Special Representatives"[5]—the distinction between the terms being a terminological one without influence on hierarchy or status[6]—are not a *novum* on the international agenda. These mandates are presently coordinated by and supported through the Office of the High Commissioner for Human Rights (OHCHR). Currently, there are 12 Special Rapporteurs and Independent Experts with a country-mandate[7] and 30 Special Rapporteurs and Independent Experts with a thematic one.[8] Country mandates are usually assigned

[3] HRC, *Human rights and access to safe drinking water and sanitation*, A/HRC/RES/7/22, adopted on 28 March 2008. While the mandate fulfills several functions, this chapter focuses on the analysis of one of its most important ones: that one of monitoring.

[4] HRC, *The human right to safe drinking water and sanitation*, A/HRC/RES/16/2, 8 April 2011.

[5] In the following the general term "Experts" will be used for "Independent Experts", "Special Rapporteurs" and "Special Representatives of the UN Secretary General".

[6] OHCHR (2001), p. 7.

[7] See http://www.ohchr.org/EN/HRBodies/SP/Pages/Countries.aspx (as of 8 January 2013).

[8] See http://www.ohchr.org/EN/HRBodies/SP/Pages/Themes.aspx (as of 8 January 2013). Additionally, there are currently six working groups focusing on specific topics.

for a period of 1 year, thematic mandates for 3 years; the maximum for a mandate to be held by the same person is usually 6 years.[9] The work of the Experts is discussed in the annual sessions of the Human Rights Council (HRC) in Geneva. Some of the Experts (depending on their mandate) also report to the UN General Assembly (UNGA) in New York.[10] The Experts' work is honorary and unpaid, but they receive personnel and logistical support from the OHCHR. This emphasizes their independence and ensures that the sole motivation for their work is their commitment to the global protection of human rights.

The experts play a key role[11] in examining specific human rights violations and analyzing how human rights principles are applied in practice.[12] It remains up to the Experts themselves to define their task and method of work in greater detail. In 1999, the Experts developed a *Manual of Operations of the Special Procedures of the Human Rights Council*, revised in 2008.[13] The manual describes the principal functions of Experts as to analyze the relevant thematic issue or country situation on behalf of the international community; to advise on the measures which should be taken by the government(s) concerned and other relevant actors; and to alert UN organs and agencies and the international community in general to the need to address specific situations and issues. Experts should also provide early warning and encourage preventive measures; advocate on behalf of the victims of violations through measures such as requesting urgent action by relevant States and calling upon governments to respond to specific allegations of human rights violations and provide redress; and activate and mobilize the international and national communities to address particular human rights issues, encouraging cooperation among governments, civil society and inter-governmental organizations.[14]

Although being different in their exact scope, what all these mandates have in common is, in one way or another, an element of monitoring in a broader sense[15]: the Experts' task is to investigate, observe and recommend solutions, where appropriate. In practical terms, this will often require investigative and diplomatic skills.[16]

---

[9] OHCHR (2001), p. 7.

[10] *Ibid.*, p. 4.

[11] Former UN Secretary General Kofi Annan called them famously the "crown jewel" of the entire HRC system, which must be preserved; see Annan (2006).

[12] OHCHR (2001), p. 4.

[13] HRC (2008).

[14] *Ibid.*, par. 5.

[15] On additional meaning and functions of the mandates, also see Steiner et al. (2008), p. 765; Hannum (2007), p. 73.

[16] For an anecdote to demonstrate these diplomatic skills, see Tullis (2010).

4.1.2 *Analysis of the Mandate's Undertaken Monitoring Efforts*

Apart from this general understanding of an Expert's role as a monitoring institution, their work is strongly dependent on the text of the resolution giving them their mandate. In addition to the general tasks mentioned above, HRC resolution 7/22 (that originally gave the Independent Expert her mandate) states three specific tasks: to further clarify the content of human rights obligations related to access to safe drinking water and sanitation; to make recommendations to help realize the Millennium Development Goals (MDGs), in particular Goal No. 7; and to prepare a compendium of good practices related to access to safe drinking water and sanitation.[17] HRC Resolution 16/2 (by which the mandate was extended and renamed) additionally encourages the Special Rapporteur, *inter alia*, to promote the full realization of the human right to safe drinking water and sanitation and to identify challenges and obstacles to this realization; to monitor the way in which the right is being realized throughout the world; and to make recommendations for goals beyond the 2015 MDG process and to facilitate, including through engagement with relevant stakeholders, the provision of technical assistance in the area of the human right to safe drinking water and sanitation.[18] In order to fulfill these tasks, the mandate chose two main tools and has thereby built on similar experiences of existing mandates: country visits and issues in focus.

4.1.2.1 Country Visits

One of the most classical means for Experts is that of conducting country visits.[19] Country visits give the mandate-holders the chance to get a first-hand impression on a given human rights situation in a country and make close contacts to governments, civil society groups and other relevant stakeholders.[20] Country visits are contingent on an invitation by that country and usually last 1–3 weeks.[21] Mrs. de Albuquerque has undertaken eleven country visits within the first 4 years: a high number given that preparation of the visit and organizing invitations are usually difficult and time-consuming.[22] In 2009, the mandate

[17] HRC, *Human rights and access to safe drinking water and sanitation*, A/HRC/RES/7/22, adopted on 28 March 2008, par. 2(a)–(c).

[18] This is a subjective selection of all the mentioned tasks (selection based on importance for this study). The rather long full enumeration of tasks can be found in HRC, *The human right to safe drinking water and sanitation*, A/HRC/RES/16/2, 8 April 2011, par. 5.

[19] HRC (2008), par. 52–74; OHCHR (2001), pp. 9–10.

[20] HRC (2008), par. 55–59.

[21] *Ibid.*

[22] Pinheiro (2003), p. 8.

conducted visits to Costa Rica, Egypt and Bangladesh.[23] In 2010, visits to Slovenia and Japan followed.[24] In 2011, the (by then re-named) Special Rapporteur visited the United States of America (USA), Namibia and Senegal followed.[25] In 2012, the Special Rapporteur conducted a visit to Uruguay and a joint visit to the pacific islands of Kiribati and Tuvalu.[26] Her first visit in 2013 was dedicated to Thailand.[27] The mandate reported on these visits to the HRC and made concrete recommendations for the specific countries. However, does the Special Rapporteur's reporting on these individual countries have the potential to make the right's implementation more effective?

Looking at the selection of the country reports, the mandate has managed to keep a geographical balance. She has visited and reported on all continents (with the exception of Australia). The mandate provides thus a truly global monitoring. It also keeps a balance in terms of the degrees to which countries are affected by

---

[23] See HRC, *Report of the independent expert on the issue of human rights obligations related to access to safe drinking water and sanitation, Catarina de Albuquerque: Mission to Costa Rica*, A/HRC/12/24/Add.1, 23 June 2009, and Official Press Statement of 27 March 2009; HRC, *Report of the independent expert on the issue of human rights obligations related to access to safe drinking water and sanitation, Catarina de Albuquerque: Mission to Egypt*, A/HRC/15/31/Add.3, 5 July 2010, and Official Press Statement of 28 June 2009; and HRC, *Joint report of the independent expert on the question of human rights and extreme poverty, Magdalena Sepúlveda Cardona, and the independent expert on the issue of human rights obligations related to access to safe drinking water and sanitation, Catarina de Albuquerque: Mission to Bangladesh*, A/HRC/15/55, 22 July 2010 and Official Press Statement of 10 December 2009. All press statements are available at http://www.ohchr.org/EN/Issues/WaterAndSanitation/SRWater/Pages/CountryVisits.aspx.

[24] See HRC, *Report of the Special Rapporteur on the human right to safe drinking water and sanitation, Catarina de Albuquerque: Mission to Slovenia*, A/HRC/18/33/Add.2, 4 July 2011, and Official Press Statement of 28 May 2010; and HRC, *Report of the Special Rapporteur on the human right to safe drinking water and sanitation, Catarina de Albuquerque: Mission to Japan*, A/HRC/18/33/Add.3, 4 July 2011, and Official Press Statement of 28 July 2010.

[25] See HRC, *Report of the Special Rapporteur on the human right to safe drinking water and sanitation, Catarina de Albuquerque: Mission to the United States of America*, A/HRC/18/33/Add.4, 2 August 2011, and Official Press Statement of 4 March 2011; HRC, *Report of the Special Rapporteur on the human right to safe drinking water and sanitation, Catarina de Albuquerque: Mission to Namibia*, A/HRC/21/42/Add.3, 28 June 2012, and Official Press Statement of 11 July 2011; and HRC, *Report of the Special Rapporteur on the human right to safe drinking water and sanitation, Catarina De Albuquerque: Mission to Senegal*, A/HRC/21/42/Add.1, 16 August 2012, and Official Press Statement of 21 November 2011.

[26] See on the visit to Uruguay HRC, *Report of the Special Rapporteur on the human right to safe drinking water and sanitation, Catarina de Albuquerque: Mission to Uruguay*, A/HRC/21/42/Add.2, 2 July 2012, and Official Press Statement of 17 February 2012; and on the visit to Tuvalu and Kiribati the Official Press Statements of 19 July and 25 July 2012.

[27] See "Press Statement of the Special Rapporteur on the human right to safe drining water and sanitation Ms. Catarina de Albuquerque" and her oral statement on the last day of the visit "Thailand: Striking contrasts in access to safe drinking water and sanitation", both of 8 February 2013. The Special Rapporteur's final report to the HRC was not yet accessible at the time of writing.

water scarcity: the Expert has visited countries with abundant water resources[28] as much as countries facing severe water scarcity.[29] The monitoring is thus neither persecution against the water-poor world nor "patting on the back" of water-rich countries.

The content of the reports represents a balance in terms of "stick" and "carrot". They appreciate countries' existing efforts for realizing human rights obligations relating to water,[30] but express critique at the same time explicitly.[31] The reports cover monitoring from the communal- to the national level: from problems of local communities[32] or discrimination against specific minorities[33] to national programs and federal legislation.[34] Although the main official addressee of the reports is the HRC, the reports are clearly also tailored towards the civil society of the visited country. The report on the visit to Japan, for instance, has also been made available in the Japanese language through the mandate,[35] so that local communities could read it. Short versions of the official reports are also made available as press releases and statements on the last day of a visit, facilitating reporting on the Expert's visit and findings in the local media in a timely manner. Finally,

---

[28] See for instance HRC, *Report of the Special Rapporteur on the human right to safe drinking water and sanitation, Catarina de Albuquerque: Mission to Slovenia*, A/HRC/18/33/Add.2, 4 July 2011, par. 10, where the expert calls Slovenia a country with water in abundance.

[29] HRC, *Report of the independent expert on the issue of human rights obligations related to access to safe drinking water and sanitation, Catarina de Albuquerque: Mission to Egypt*, A/HRC/15/31/Add.3, 5 July 2010, par. 10 states that the country faces water scarcity.

[30] See, for instance, HRC, *Report of the independent expert on the issue of human rights obligations related to access to safe drinking water and sanitation, Catarina de Albuquerque: Mission to Costa Rica*, A/HRC/12/24/Add.1, 23 June 2009, section "Good Practices", par. 54–60; or HRC, *Report of the independent expert on the issue of human rights obligations related to access to safe drinking water and sanitation, Catarina de Albuquerque: Mission to Egypt*, A/HRC/15/31/Add.3, 5 July 2010, section "Achievements", par. 6–9.

[31] See for instance HRC, *Report of the independent expert on the issue of human rights obligations related to access to safe drinking water and sanitation, Catarina de Albuquerque: Mission to Costa Rica*, A/HRC/12/24/Add.1, 23 June 2009, section "Issues of Concern", par. 61–73; or HRC, *Report of the independent expert on the issue of human rights obligations related to access to safe drinking water and sanitation, Catarina de Albuquerque: Mission to Egypt*, A/HRC/15/31/Add.3, 5 July 2010, section "Challenges", par. 10–12; Statement on last day of visit to Thailand, par. 2 ff.

[32] See for instance HRC, *Report of the independent expert on the issue of human rights obligations related to access to safe drinking water and sanitation, Catarina de Albuquerque: Mission to Costa Rica*, A/HRC/12/24/Add.1, 23 June 2009, on several visited provinces, par. 5 and 46–48.

[33] See for instance HRC, *Report of the Special Rapporteur on the human right to safe drinking water and sanitation, Catarina de Albuquerque: Mission to Slovenia*, A/HRC/18/33/Add.2, 4 July 2011, section on "Roma", par. 17–19.

[34] See for instance HRC, *Report of the independent expert on the issue of human rights obligations related to access to safe drinking water and sanitation, Catarina de Albuquerque: Mission to Costa Rica*, A/HRC/12/24/Add.1, 23 June 2009, on the national program for potable water improvement and quality sustainability of potable water services and the "Seal of Quality Public Health Program" and the "Blue Flag Ecological Program", par. 58–59.

[35] See report available in Japanese language at http://www2.ohchr.org/english/issues/water/iexpert/docs/Press_statement_japaneseJuly2010.doc.

the Expert has put human rights obligations related to water access in specific countries in the context of other human rights: where appropriate, she conducted country visits together with other Experts.[36]

In sum, the monitoring of individual countries' track record with regard to the fulfillment of the human right to water is noteworthy. The reports on the respective countries are far-reaching and well-balanced. By choosing exemplary countries and highlighting good and bad practices alike, both local and national, the mandate manages to provide an excellent overall overview over compliance with and violations of human rights related to water on a global scale.

4.1.2.2 Issues in Focus

The second main tool chosen by the mandate is that of "issues in focus". These special focuses are meant to direct special attention to one particular subtopic of a mandate. Experts dedicate usually up to one calendar year of their work to a special topic, and file corresponding reports. The Special Rapporteur has chosen several "issues in focus", namely "sanitation" in 2009,[37] "private sector participation" in 2010,[38] "stigmatization" in 2012[39] and "sustainability and non-retrogression" in 2013.[40] Additionally, the mandate dedicated special attention to the issue of the "contribution of the human right to water and sanitation to the MDGs",[41] (recently followed-up by a "post-2015 development agenda for water, sanitation and hygiene"),[42] and to the issue of "climate change and the human right to water and

[36] See in particular HRC, *Joint report of the independent expert on the question of human rights and extreme poverty, Magdalena Sepúlveda Cardona, and the independent expert on the issue of human rights obligations related to access to safe drinking water and sanitation, Catarina de Albuquerque: Mission to Bangladesh*, A/HRC/15/55, 22 July 2010, which the expert conducted together with the Independent Expert on the question of human rights and extreme poverty, Ms. Magdalena Sepúlveda Cardona.

[37] See HRC, *Report of the independent expert on the issue of human rights obligations related to access to safe drinking water and sanitation, Catarina de Albuquerque*, A/HRC/12/24, 1 July 2009.

[38] See HRC, *Report of the independent expert on the issue of human rights obligations related to access to safe drinking water and sanitation*, A/HRC/15/31, 29 June 2010.

[39] See HRC, *Stigma and the realization of the human rights to water and sanitation*, A/HRC/21/42, 2 July 2012.

[40] Report not yet available. For further information, see http://www.ohchr.org/EN/Issues/WaterAndSanitation/SRWater/Pages/Sustainabilityretrogressionrealisationofwaterandsanitation.aspx.

[41] See UNGA, *Human rights obligations related to access to safe drinking water and sanitation*, A/65/254, 6 August 2010.

[42] See UNGA, *Integrating non-discrimination and equality into the post-2015 development agenda for water, sanitation and hygiene*, A/67/270, 8 August 2012.

sanitation".[43] The mandate has furthermore identified a compendium of "good practices"[44] and is currently developing a comprehensive "Handbook for Implementing the Human Rights to Water and Sanitation: From Policy to Practice“.[45]

Analyzing the mandate's choices for these special topics, many of the thematic focuses emphasize how seriously the Expert takes the tasks as laid down in resolution 7/22 (her first mandate) and resolution 16/2 (the extension of the mandate). Focuses such as the compilation of good practices, the connection with the MDGs or the focus on vulnerable groups (stigmatization) mirror exactly the wording of the two resolutions that gave the Expert her mandate. This way the mandate protects itself from being accused of going beyond its approved tasks or of having an own political agenda.

The additional thematic focuses that the mandate chose emphasize that the Expert understands monitoring not only as being a watchdog over violations of the human right to water, but also as the chance to facilitate the implementation of the right: how can positive experiences in some communities be made accessible to others, and how can existing efforts be made more sustainable? With respect to the working method, the Special Rapporteur regularly holds expert consultations and public consultations to approach a new topic and makes written input from all stakeholders possible.[46] The mandate thus attempts to include a variety of different voices and experiences.

Analyzing these issues in focus, it becomes clear that the mandate has applied two strategies for its choices of the respective topics: to choose either a "controversial topic" (a sub-question of the mandate that is highly contested between different stakeholders) or to direct attention to a "neglected issue" (a question which has been omitted in prior debates).[47] Issues of controversy are those topics identified a particularly important on which the mandate is expected to take a clear stance; a neglected issue is one on which new debate is needed, as it would for the first time prominently be brought to the forefront of the agenda. In that sense, given the large

[43] Position paper available at www.ohchr.org/Documents//Issues/Water/Climate_Change_Right_Water_Sanitation.pdf.

[44] See HRC, *Report of the Special Rapporteur on the human right to safe drinking water and sanitation, Catarina de Albuquerque: Compilation of good practices*, A/HRC/18/33/Add.1, 29 June 2011.

[45] See announcement of the book and the quest for public input at http://www.ohchr.org/EN/Issues/WaterAndSanitation/SRWater/Pages/FromPolicyToPractice.aspx.

[46] See for instance the stakeholder consultations on "sanitation" at http://www.ohchr.org/EN/Issues/WaterAndSanitation/SRWater/Pages/SanitationConsultation.aspx, or the written responses of stakeholders on the issue of "stigmatization" at http://www.ohchr.org/EN/Issues/WaterAndSanitation/SRWater/Pages/Contributionsstigmatization.aspx.

[47] This is not official terminology employed by the mandate. The author uses these terms to capture the essence of these two approaches.

civil society movement against water privatization[48] and the huge financial interest that water companies have in private water investment, private sector participation certainly is the most controversial issue in the field of water and was thus a logical choice for the mandate. Sanitation and stigmatization, on the other hand, have been neglected issues, as they can often be a less comfortable topic to talk about.

Both approaches have distinct advantages: in the case of a "neglected issue", the impact of an issue in focus is particularly high; in the case of a "controversial topic", the positioning of the mandate, as an authoritative and independent body, is particularly important: the mandate simply cannot be "silent" about such topics. The choice of controversial topics as well as neglected issues emphasizes that the mandate has chosen a well-balanced approach to its monitoring efforts. The methodology employed by the mandate furthermore emphasizes an inclusive approach: public input is made possible on the large scale from individuals and civil society; regular stakeholder meetings are held in order to hear the interested groups and bring them together. In combining all these experiences from different regions and different countries, the Expert's work is another good example that the right to water combines a multitude of different rights and approaches rather establishing one uniform standard.

4.1.3 Conclusion and Remaining Problems

Generally, the assessment of the monitoring undertaken by the Special Rapporteur must be very positive. It is in particular balanced along two axes.

Firstly, the mandate focuses on both water-scarce and water-abundant countries for its case studies. This way, it is credibly neither a public prosecutor of the group of water-scarce States nor the messenger for those States that are water-abundant. Secondly, the mandate concentrates on both controversial and neglected topics: in the one case, the mandate has an impact in influencing the already on-going discussion and strengthens the human rights approach in that debate; in the other case, the mandate brings certain topics to the agenda in the first place.

However, such a positive outlook on existing monitoring efforts would be naïve, if it did not also point to existing problems, which need further consideration for the monitoring to become even more effective. Firstly, the mandate, like any mandate of an Expert, is dependent on getting invitations for country visits, as any other practice would be a clear violation of Art. 2 No. 7 UN Charter. However, this way countries that might be most severely in breach of their human rights obligations, or in which disputes over water are most pressing, remain excluded from the direct monitoring mechanism of country visits as long as

[48] See, most prominently, the European campaign "Water is a Human Right" with far over 1 million online signatures (as at 20 February 2013), see http://www.right2water.eu. See also Hall et al. (2005).

they refrain from inviting the Special Rapporteur. To name an example, a visit of the Special Rapporteur to Israel and the Palestine Territories might appear amongst the most important ones, but is at the same time most unrealistic, as water remains a particular explosive topic in that area. However, the visit to the USA in 2011 shows that also countries traditionally opposed to the right to water (as to any socio-economic rights) have started to cooperate with the mandate and have opened their gates for country visits.

Secondly, the Special Rapporteur is bound strictly to the wording of her mandate. Should the HRC change the underlying resolution substantially, for instance by banning a particular topic from her monitoring, as happened in the case of international watercourse law and transboundary water issues,[49] the Rapporteur's hands remain bound. The Special Rapporteur's mandate is a result of negotiations in the HRC and thus a compromise between different positions of Member States. Thus, the monitoring of the right to water through the Special Rapporteur can only be as comprehensive as the mandate allows, and this mandate is designed by Member States.

Thirdly, Experts and commentators have repeatedly pointed out that the massive understaffing of the mandates impedes on the effectiveness of their work.[50] Abdefattah Amor, former Special Rapporteur on the elimination of all forms of discrimination and intolerance based on religion or belief, famously stated that "the human rights section of the United Nations is the third world of it"[51]: human and material resources for Experts are entirely insufficient. The Special Rapporteur on the human right to water might have been successful in raising funds from third parties (thanks to the "hyped" topic of water). To rely on external sources, such as fundraising from the private sector, is certainly an inevitable option, as long as it does not put the Experts' independence. Connecting with independent research institutions for support in logistics and human resources, as the Special Rapporteur has done,[52] is a preferable option, but is not always feasible or sufficient.

---

[49] HRC, *Human rights and access to safe drinking water and sanitation*, A/HRC/RES/7/22, adopted on 28 March 2008, the preamble affirms "the need to focus on local and national perspectives in considering the issue", but leaves aside "questions of international watercourse law and all transboundary water issues".

[50] Pinheiro (2003), p. 8; Amor (1998), p. 947; see also Centre for International Governance (2010), in particular contributions by T. J. Piccone (p. 9) and P. O' Connell (p. 18).

[51] Amor (1998), p. 947.

[52] The Special Rapporteur, for instance, has connected with the German Institute for Human Rights ("Deutsches Institut für Menschenrechte"), which is an independent association under German law and sees itself as an independent institution of civil society; see the Institute's Mission Statement, http://www.institut-fuer-menschenrechte.de/en/about-us/mission.html. The institute supports the mandate by financing an external researcher, employed at the Institute. The Special Rapporteur also connects with other research institutions on a case-by-case basis, see for instance Position Paper on "Climate Change and the Human Right to Water", http://www.ohchr.org/Documents/Issues/Water/Climate_Change_Right_Water_Sanitation.pdf, Fn. 1, about the cooperation with the Swedish International Development Cooperation Agency.

A final problem remains that of lacking follow-up mechanisms after the country visits[53]: once a visit is conducted and the report is released to the HRC, what are the available mechanisms of follow-up to ensure that the necessary changes are implemented? These mechanisms are rare, if existent at all. The Universal Periodic Review (UPR)[54] of the HRC is a first step in this direction: it involves a review of the human rights records of all UN Member States once every 4 years. However, this review is rather broad and not tailored specifically for the follow-up of reports of Special Rapporteurs. Given the mentioned constraints in finances and staff, the mandate will also not be able to provide follow-up mechanisms herself. It can only be a facilitator for other follow-up mechanisms, by means of schooling and informing civil society groups and research centres in the countries they have visited.

Altogether, apart from these caveats, the mandate of the Special Rapporteur provides monitoring of the right to water of high quality. Its distinct quality lies in its well-balanced comprehensiveness: it covers water-abundant and water- scarce countries; it deals with neglected issues as well as those that are most contested; it identifies negative examples ("violations") and positive examples ("good practices") alike; it includes views of civil society and all relevant stakeholders; its addresses its reports to the UN, but at the same time to governments, to the media and to civil society. Thus, independent monitoring for the right to water, certainly one of the crucial preconditions to make the right's implementation effective, is well in place.

4.2 Enforcing the Right in the Private Sector: The Case of Water Privatization

In order to be effective, the right must also be enforceable not only against public service providers, but also against the private ones. This chapter examines the special case of the right to water in times of water privatization. In this scenario, the right to water is often challenged: people's collective right to water can easily clash with the rights of investors. Could a right to water, if we accept in the ways suggested in Chaps. 2 and 3, be enforced against the private sector, even where it clashes with legitimate rights of investors?[55]

---

[53] See also Final Recommendation 8 of Centre for International Governance (2010), which also identifies this as a mayor problem, p. 20.

[54] UNGA, *Human Rights Council*, A/RES/60/251, 3 April 2006, which created the HRC firstly mentions the UPR in par. 5e). It was further elaborated in HRC, *Institution building of the United Nations Human Rights Council*, A/HRC/RES/5/1, 7 August 2007, Annex.

[55] I have given an extended analysis of this topic at Thielbörger (2009), pp. 487–510.

4.2.1 *Preface: Conceptual Problems with the Right to Water and Privatization*

Recalling the essential history and background of water privatization, which is a wide-spread phenomenon in the water sector already since the 1980's, when Margaret Thatcher privatized Britain's water systems in the most radical way: by selling the existing infrastructure to private parties.[56] Water privatization is nowadays a global movement; several different forms can be distinguished.[57] Apart from many developed countries, also many threshold and developing countries decided to privatize their water supply systems over the last decades. Two reasons are in particular identifiable: firstly, privatization is often a tool to improve water supply through private capital that otherwise would be out of reach; secondly, the World Bank often makes additional development aid for developing States conditional upon the privatization of their water supply systems. The question is: to what extent is water privatization compatible with the concept of the right to water, and how must water privatization be designed in order to match the requirements set out under this human right?

The first reaction to this question might be: why would privatization not be compatible with the right to water at all? Following Adam Smith's famous idea of the "invisible hand",[58] the assumption of mainstream economics[59] is that, under normal circumstances, the marketplace regulates itself: when, in a given market, demand equals supply, overall social surplus is maximized, so the basic model goes. Consumers and suppliers combined are best off. At this point (the so called "market equilibrium"), a good is traded at a certain price and is only produced to that amount which will entirely be consumed. This way, there is neither over- nor underproduction.[60]

Opening water supply systems for the competitive market must therefore be, generally speaking, a welcome idea. If water is like any other good or service, the market mechanisms aim to maximize the overall social surplus.[61] However, as I will argue in this chapter, privatization in the case of water is a double-edged sword: along with the envisaged benefits, certain modes of privatization of water services

---

[56] The most relevant law in this respect is the UK Water Act of 1989. On this topic, see also Fischer and Zwetkow (2003), p. 133; Irmscher (2001), p. 160; and Laskowski (2010), pp. 657–670.

[57] See for a good overview, Riedel (2006), pp. 21–22, and Laskowski (2010), pp. 838–841.

[58] Smith (1790), Part IV ("Of the Effect of Utility upon the Sentiment of Approbation"), at I.10.

[59] To distinguish this approach, which dominates academia, politics and media alike, from other economic approaches, with regard to the human right to water, see Branco and Henriques (2010), p. 143.

[60] Out of many to explain the general assumption of the basic model of mainstream economics: Pindyck and Rubinfeld (2009), p. 25.

[61] It is a very different question, of course, whether the market solution is the *socially* most desirable outcome, thus whether it distributes the surpluses adequately between consumers and suppliers.

carry significant risks and caveats. This has even led some authors in the past to argue that water privatization can never succeed as it creates unsolvable problems[62] or that it can only be undertaken under very strict regulation of the private supplier.[63] So what are those assumptions under mainstream economics that make the concept of a human right to water so difficult to reconcile with the idea of market principles?[64]

4.2.1.1 Market Principles Incompatible with Services to Fulfill Basic Needs?[65]

Nobel-prize laureate Amartya Sen famously pointed out that rights, as a concept, are not *per se* important for traditional economic analysis.[66] Taking this thought one step further, are rights to fulfil basic needs under all circumstances even *compatible* with the idea of the market? Can a human right to water and the market principles conceptually go together?

First of all, the general idea in traditional economic analysis—that the market equilibrium is the desirable outcome in a given market—does not hold for the case of rights to ensure the fulfilment of basic needs (of which the right to water is certainly one). At the market equilibrium, not all potential water consumers have access to water: only those who are willing or able to pay the market price are consumers in the market.[67] In turn, poor or vulnerable households that are not able to pay the market price of water are left out. Such a result is incompatible with the ideal of universal coverage, which is the only logical aim of a human rights approach.[68] Thus, if a service fulfils a vital need, as water supply does, the aim in a given market is regularly not to reach the market equilibrium. The aim must rather be to serve the highest quantity of consumers possible (as long as the used water amount replenishes itself and is not damaged in the long run).[69] This, however, would put the supplier in turn in an impossible situation: suppliers would need to sell water far below the market price, which would mean economic losses for them. This dilemma emphasizes that, without State intervention, a human right to water in order to fulfil basic needs can be in logical conflict with the basic principles of market

[62] See Lobina (2000).

[63] See for instance Beisheim (2007), pp. 109–119; Dalton (2001).

[64] Branco and Henriques (2010) and Jeffords and Shah (2012) come to similar caveats, albeit in a more quantitative manner. What I offer here, is an conceptual and non-quantitative account.

[65] This chapter is not a in-depth economic analysis. It will highlight those basic economic assumptions that the human right to water is not easily compatible with. For an economic in-depth analysis of how to merge economic analysis with the framework of the human right to water, see Jeffords and Shah (2012).

[66] Sen (1987), p. 47.

[67] Pindyck and Rubinfeld (2009), p. 25.

[68] Branco and Henriques (2010), p. 150; Jeffords and Shah (2012), p. 26.

[69] Jeffords and Shah (2012), p. 27.

economy: human rights must either be guaranteed for all rights-holders (at least for as many as currently possible, and that number must be steadily increased) or the guarantee goes against the very idea of a human right. In this way, the best outcome in traditional economic analysis can easily be a violation of human rights principles, in particular the idea of universality, the special protection of vulnerable groups or the requirement to supply a minimum amount of water at all times.

4.2.1.2 Non-existence of Competitive Water Markets?

Even if we disregard this intrinsic conflict between the market principle and the logic of a human right to water, water markets will nevertheless likely incur failures. In particular, the introduced model of traditional economic analysis is based on the idea of competitive markets. Thus, in order to produce the desired results, not only a rather big number of consumers is required, but also a variety of suppliers in a given market.[70]

This assumption, however, does not hold in typical water markets, where water is serviced in the form of tap water. The situation more often resembles that of a natural monopoly[71]: Albeit there is a big number of consumers, supply is usually provided only by one firm alone ("monopoly"). For water specifically, fixed costs (costs that exist independent of the overall amount of water delivered) largely outweigh marginal costs (the costs of every additional unit of water): establishing a new delivery network is far too costly for new competitors to enter the market, so that the first (or strongest) supplier keeps the market for itself. For technological reasons, the infrastructure in the case of water delivery cannot be shared; it is run and used by one supplier alone. Water consumers, different from consumers of energy or telephone, therefore can usually not choose between different suppliers. If competition exists at all, it is in the form of competition to get onto the market. In these cases, as we will see in the following case studies, different water suppliers make bids for exclusive concession rights. Once on the market, however, they hold monopolies. Water suppliers therefore often have market power[72]: they are not price-takers as they do not have to accept the dominating price in the market. They rather dominate the rules of the game themselves: the monopolist, so mainstream economics assumes, chooses quantity of delivered units so that it maximizes its own profits, not the profits of both suppliers and consumers combined.[73] This fatal combination entails a higher price than the equilibrium price and a lower supplied overall quantity than would be socially optimal. This, of course, not only jeopardizes the realization of universal access to water; it even prevents the market to deliver the optimal social surplus so that consumers and suppliers combined would be best off.

---

[70] Pindyck and Rubinfeld (2009), p. 271.

[71] Branco and Henriques (2010), p. 153; Lever (2007), p. 87; Laskowski (2010), p. 664 (using the example of the United Kingdom (UK) to explain the transition from public to private monopolies).

[72] Pindyck and Rubinfeld (2009), p. 349.

[73] *Ibid.*, pp. 391 ff.

This caveat—the lack of competition in a give water market—however, holds only true for tap water; in the case of bottled water, competition is generally possible (but dependent on the individual case, in particular the physical accessibility of a given market for a high number of small vendors).

4.2.1.3 Water as a Non-private Good?

Economic analysis usually distinguishes between two main categories of goods: public and private goods.[74] Assumptions along the introduced model of mainstream economics are mainly tailored towards private goods. However, can water easily be categorized as such a private good, even if we accept the existence of a human right to water?

In traditional economic analysis, a public good, regardless of who holds the moral ownership,[75] is a good that is non-rivallous and non-excludable. Non-rivalry of a good or service means that its consumption by one individual does not reduce the amount left for others to be consumed (so that the marginal cost of its provision to an additional consumer is zero); non-excludability of a good or service is assumed, if, in practical terms, no consumer can effectively be prevented from consuming that good or service, so that charging for the individual use is difficult.[76] The classical example for a rival and excludable, and thus private, good is any particular piece of food: once consumed, it cannot be eaten by others anymore, and by pure physical means individuals can be prevented from accessing it. Air, on the other hand, is a typical public good: breathing it does not (significantly) reduce the overall amount of air available to all humans, and, given air's omnipresence, it is very difficult to prevent individuals from breathing it. Intuitively, we feel that drinking water is in some ways like food; in some other ways, however, we might feel that water is also similar to air. This paradox emphasizes already that the categorization of water as public or private good is difficult.

Accepting the two parameters of "excludability" and "rivalry" for the classification of goods, there are two more kinds of goods thinkable: those that are rivallous, but non-excludable (so called "common pool goods") and those that are non-rivallous, but excludable (so called "toll goods").[77] Which of these four categories would water fall into—is it a public, private, common pool or a toll good? Only if water is a private good, could one apply the rules of mainstream economics without further adjustment.

---

[74] *Ibid.*, p. 676.

[75] On this issue, see the notion of common ownership of water in Sect. 4.3.1.5 of this book.

[76] Pindyck and Rubinfeld (2009), p. 676. Note that the economic language differs from political language, which often declares water a "public" or "common" good in order to emphasize that it morally belongs to the public and that its use must reflect social needs. This political statement, however, has little to do with the economic classification. On this, see also Branco and Henriques (2010), p. 144.

[77] Pindyck and Rubinfeld (2009), pp. 676 ff.

In order to answer this question, one has to distinguish between the possible sources of the drinking water in question. Natural water of rivers and lakes as well as rainwater, used as main basic natural drinking water source by many indigenous people, is most clearly not a private good. Whether such waters are "rivallous", is already in doubt: they are regularly, and without human effort, renewed through periodic rains. However, Garrett Hardin has shown in his famous example of "The Tragedy of The Commons"[78] that even a renewable good, if used by many individuals of a community, is likely to be over-used and therefore damaged in the long run, unless property rights and distributive rules are introduced. This "tragedy" is also true for many natural water bodies which puts their non-rival character in doubt: using too much of it, excludes others from doing the same. The rivallous character of natural waters is thus contestable. However, natural waters are at least clearly non-excludable: it is hard to isolate the whole of water bodies (almost impossible in the case of rain water). Thus, natural water sources are either public goods (if one assumes, they replenish themselves, so that non-rivalry occurs) or they are at least common pool goods.[79] Applying traditional economic analysis to these kinds of water sources is problematic: mainstream economics is mainly tailored to the logic of private goods.

However, advanced access to tap water is a different case. If a household receives water from the tap, this water is, economically speaking, a private good: on the one hand, water access for individual households can physically be disconnected by the supplier as is often done in case of continued non-payment ("excludability"); on the other hand, treated water within the supply system is finite and must be renewed under human effort ("rivalry").[80] From an economic point of view, tap water is excludable and rivallous, and thus a private good.

This *prima facie* finding, however, appears dissatisfying from a human rights perspective. Considering tap water a private good in economic terms neglects that under a human right to water a minimum water supply must, under all circumstances, be ensured. Disconnection—to *exclude* individuals from basic water access—is considered a violation of their human rights by most legal texts and by most judges. Thus, classifying water as a private good, as traditional economic analysis would suggest, means allowing a behaviour, namely disconnection, which regularly constitutes a violation of the right. Thus, although seeing tap water as a private good is consistent with the model of mainstream economics, a human right to water is much harder to reconcile with the idea of a private good—at least, if the market supply is not adjusted to certain indispensible human rights requirements. These must include ensuring a minimum supply of water at all times and granting specific protection to vulnerable groups (e.g. connecting rural populations to water supply that are often excluded from accessing treated water due to higher costs of connecting rural areas to the infrastructure).

---

[78] Hardin (1968), pp. 1243 ff.

[79] Branco and Henriques (2010) categorize natural waters rather as the latter, p. 145.

[80] *Ibid.*, p. 145.

4.2.1.4 Current Dynamics of Supply and Demand: Increasing Water Prices?

So, what would happen if water was left to the private markets (disregarding the previous three conceptual caveats for a moment)? Unless effective water policies and technological innovations are put in place, the global demand curve for water will keep shifting outwards, as the global population is still increasing[81]: more water is needed in the future. On the other hand, if overuse of water continues, the global supply curve of water will shift inwards: less water will be available. As the only logical result, the equilibrium price of water must rise, leading to even more social injustice as poor households can even less afford to pay for the water they need at the price the water demands.

4.2.1.5 Conclusion: Compatibility of the Right to Water and Privatization

Should this invoke the conclusion that a human right to water and water privatization are generally incompatible? Such an assumption would be throwing out the baby with the bathwater. General Comment No. 15[82] as well as the 2010 report of the Special Rapporteur on "Private Sector Participation"[83] make clear that privatization of water is *not* generally incompatible with the idea of a human right. The outreach to private money and expertise is one means to realize the right to water. However, the outlined potential problems have shown that privatization must be approached with great caution.

Firstly, to achieve the market equilibrium cannot be the goal in the case of privatising water delivery; the goal, which the human right to water clearly demands, must be to come closer to universal coverage. Secondly, States have to take special account of the fact that water markets are regularly natural monopolies: they need to think of ways of how to balance the private suppliers' market power in order to protect the vulnerable consumers. Thirdly, even if exposed to the private market, tap water cannot be treated as a mere private good. Although common economic classification suggests to do so, qualifying tap water as a private good would imply the possibility to exclude individuals from accessing water. This, however, would in turn mean allowing for a violation of the human right to water. Thus, water privatization must be done with a view to ensuring basic access also in the case of continued non-payment. Finally, price increases are the inevitable outcome of leaving water to the market, if not counterbalanced through State

[81] See the overview of the UN Department of Economic and Social Affairs on water scarcity, http://www.un.org/waterforlifedecade/scarcity.shtml.

[82] ESCR-Committee, *General Comment No. 15*, par. 27 and 49.

[83] HRC, *Report of the independent expert on the issue of human rights obligations related to access to safe drinking water and sanitation*, A/HRC/15/31, 29 June 2010.

control and mandatory technological investments through the State and the private investor. Where water demand increases, due to a growing population, and water supply decreases, due to chronic overuse, the only logical economic consequence is price increases—a development that is not compatible with the ideal of universal coverage under the human right to water.

4.2.2 Case Studies: The ICSID-Jurisdiction on Water Privatization

With these economic parameters in mind, this chapter continues to examine the way cases of water privatizations have been legally decided in the past.[84] One of the most important institutions to hold jurisdiction over disputes following failed water privatizations is the International Centre for the Settlement of Investment Disputes (ICSID).[85] As an arbitrational tribunal situated within the framework of the World Bank, the ICSID settles disputes between host States and foreign investors of all kinds. Whether and to what extent a right to water has been taken into account by States, private investors and the ICSID, will be examined in the following sections. Has the emerging right to water made a difference in the settlement of these disputes, and could it do so in the future, if even more clearly accepted as a human right? The three following case studies will provide a starting point in answering this question.

4.2.2.1 Compania de Aguas de Aconquija v. Argentine Republic[86]

In the province of Tucumán, Argentina, the local government decided in 1993 to privatize its water and sewage facilities, due to pressure onto public finances in the province. Only one consortium, led by the Compagnie Générale des Eaux (CGE), a French corporation that later became Vivendi Universal, made a bid. Although the bid did not meet the certain requirements—it included a reduced amount of mandatory private investments and higher prices for customers than suggested by the provincial authorities—the province of Tucumán accepted a 30 years water concession with CGE's Argentine affiliate Compania de Aguas del Aconquija (AdA), in which the company committed itself to invest US$384.5 million throughout the time of the concession.

[84] I have offered a similar analysis of these three cases at Thielbörger (2009), pp. 493–503.

[85] Reed et al. (2011), pp. 1–22. See also the overview "About ICSID", http://icsid.worldbank.org.

[86] For the facts and the procedural history, see ICSID, *Compañía de Aguas del Aconquija, S.A. & Compagnie Générale des Eaux v. Argentine Republic*, Case No. ARB/97/3, Award of 21 November 2000, sections A–C. Elements of this description and analysis have been drawn from my own previous work, see Thielbörger (2009), pp. 493 ff.

After two instances of water turbulence, which had turned Tucumán's watercolour temporarily red and black, and a public upheaval in 1996, the newly elected governor re-negotiated the contract, seeking to reduce prices for customers and higher mandatory investments through AdA. Several officials encouraged the population not to pay their water bills. After a compromise had seemingly been found, the governor of Tucumán, however, submitted a substantively altered contract to the legislature, which AdA eventually refused to sign. In 1997, CGE and AdA notified the rescission of the concession contract, and filed a claim with the ICSID against the Argentine Republic. It claimed Argentina's liability under the France-Argentina Bilateral Investment Treaty (BIT) and the Convention on the Settlement of Investment Disputes between States and Nationals of other States (ICSID-Convention), seeking in total US$300 million for damages.

In the subsequent award, the arbitrators were mainly concerned with two separate legal questions: firstly, whether the tribunal had jurisdiction over the dispute at all, and, if so, secondly, whether Argentina should be held liable. Only the first question was answered positively; on substantial grounds, however, the claim was dismissed. In the subsequent annulment procedure of that award, the Annulment Committee found that the original tribunal, in failing to decide whether there had been a breach of the BIT, committed a "*manifest excess of jurisdictio*n" and thus overruled the original award.

Relevant Reasoning of the Tribunal

In the original award, the arbitrators firstly examined whether they had jurisdiction. It acknowledged that the parties had agreed in the contract to exclusively submit any arising dispute to the local courts of Tucumán.[87] However, the tribunal also noted that under Art. 25 ICSID-Convention and under Art. 8 of the French-Argentine BIT, its jurisdiction extended to *any legal dispute* arising out of an investment by a French investor in Argentina.[88] Relying on these provisions, the tribunal assumed its own jurisdiction.

On the substantial issue of Argentina's responsibility, the claimants had argued that some of Tucumán's actions had resulted in a fall in the recovery rate under the concession contract, in the unilateral reduction of the tariff rate and that provincial officials had altogether acted in bad faith.[89] The tribunal, to the contrary, found that Argentina could not be held liable under the BIT until the companies had tried to enforce their rights through the local courts of Tucuman, because a detailed interpretation and application of the concession contract would be necessary to

[87] ICSID, *Compañía de Aguas del Aconquija, S.A. & Compagnie Générale des Eaux v. Argentine Republic*, Case No. ARB/97/3, Award of 21 November 2000, p. 2, referring to Art. 16.4 of the concession contract.

[88] *Ibid.*, par. 45.

[89] *Ibid.*, par. 63.

decide on the alleged violations of the BIT.[90] This right, however, had been attributed in the contract exclusively to the jurisdiction of the Tucumán courts.

Following this dismissal, the claimants sought partial annulment of the award, arguing in particular that the tribunal had "manifestly exceeded its powers".[91] The Annulment Committee stated that the tribunal—contrary to its own assumption—did in fact have the jurisdiction to base its decision upon the concession contract. It confirmed that a "State may breach a treaty without breaching a contract, or vice versa".[92] The tribunal of the original award should thus have examined whether the provincial authorities had breached their obligation to grant "fair and equitable treatment"[93] as demanded under Art. 3 of the BIT. If this analysis made an implicit interpretation of the concession contract necessary, the tribunal should have carried out that task which the tribunal, however, had refused to do. CGE's and ADA's annulment procedure was thus eventually successful.

Assessment of the Awards in Light of the Human Right to Water

One should firstly acknowledge that an actual water-crisis never appeared in Tucumán. In particular, experts did not consider that any of the two turbulences that changed the colour of Tucumán's water for a while posed any health risk to the population.[94]

However, it is noteworthy that a violation of the right to water can occur already by concluding a concession contract that is in (abstract rather than concrete) conflict with the right to water: although the Tucumán officials had originally realized the need to guarantee stability of water prices and quality also after the privatization, they finally decided to accept a bid which would not ensure these standards. This is somewhat understandable given that the government had only one bidder to negotiate with. However, it is nevertheless a violation of the right to water: in accepting a flawed concession agreement, States manoeuvre themselves into a dilemma, namely the inevitable clash between their popular obligation under the right to water and their self-made obligations towards the investor under the concession contract. This dilemma led the governor of Tucumán to try and force AdA to renegotiate the agreement. Even crediting him with good intentions towards the right to water, the governor thereby undoubtedly violated at the same time the rights of AdA established through the concession contract. In order to avoid dilemmas of this kind, States must, already during the very first phases of negotiating the concession contract, keep their obligations with respect to a human right to water in mind. If no

[90] *Ibid.*, par. 79.

[91] ICSID, *Compañía de Aguas del Aconquija, S.A. & Compagnie Générale des Eaux v. Argentine Republic*, Case No. ARB/97/3, Decision on Annulment of 3 July 2002, par. 2.

[92] *Ibid.*, par. 95.

[93] See on this provision in general: Schill (2006); Knoll-Tudor (2009).

[94] See quote at Mayeda (2008), p. 194.

bidder is willing to accept the required guarantees of water quality and water prices as demanded under a right to water, privatization of water services must not proceed; the State remains responsible alone.

Secondly, the most remarkable part of the AdA award might be the issue of jurisdiction. As Thomas Wälde points out, "[f]oreign investors tend to be successful in international arbitration, mainly because the law (BIT, contract) is largely on their side".[95] Thus, it is somewhat understandable that provinces like Tucumán have tried to grant exclusive jurisdiction over the concession contract to their local courts instead of accepting ICSID arbitration.[96] However, given the AdA case, it will be hard, if possible at all, for States in the future to bypass ICSID arbitration. The Annulment Committee has made its point clear that it will not accept States' attempts to deprive the ICSID tribunals of their jurisdiction, by granting exclusive jurisdiction over the interpretation of the concession contract to local courts. It will be even more important for States in the future, rather than engaging in ultimately unsuccessful attempts to bypass ICSID arbitration, to make sure to protect the right to water in their own actions from the very beginning. States will not be able to rely on the favorable jurisprudence of their own local courts: the ICSID tribunals remain the crucial instance for the settlement of water investment disputes in the future and, if necessary, the arbitrators are prepared to interpret the concession contracts.

4.2.2.2 Azurix Corporation v. Argentine Republic[97]

In 1999, US-based water services firm Azurix, an Enron spin-off, paid US$438.6 million in an auction for an exclusive 30 years-concession to take over the water and sewage systems in the Province of Buenos Aires. In March 2000, conflicts about bad water quality, including algae, and water pressure arose. Authorities warned half a million residents that their local water supply had been intoxicated by bacteria. Customers were advised to boil their tap water, to avoid drinking the local water, and not to pay their water bills. Local authorities talked about the biggest water crisis in the province for at least 25 years.[98] Azurix countered that the Province of Buenos Aires had agreed in the concession contract to complete certain repairs of the equipment and systems, which was never done. Azurix also claimed that the Province had attempted to interfere with the water tariffs and had purposefully incited public panic. The company terminated the concession contract in

[95] Wälde (2004), p. 1135.

[96] Another argument commonly employed by States in favor of the jurisdiction of local courts is, of course, the bigger familiarity of local courts with the communal conditions and needs.

[97] For the facts and the procedural history, see ICSID, *Azurix Corporation v. Argentine Republic*, Case No. ARB/01/12, Award of 14 July 2006, sections II, III and VI. Elements of this description and analysis have been drawn from my own previous work, see Thielbörger (2009), pp. 496 ff.

[98] Perin (2000).

October 2001 and subsequently filed a claim at the ICSID under the US-Argentina BIT. The company claimed more than US$550 million in compensation, alleging violations of the guarantees concerning expropriation, fair and equitable treatment and security and protection. After a preliminary decision on jurisdiction in December 2003, the ICSID arbitrators granted compensation to Azurix: however, this compensation amounted only to US$165 million (less than a third of the originally sought compensation). The following annulment procedure ended with the "Decision on the Continued Stay of Enforcement" in December 2007, which largely confirmed the findings of the original award.

Relevant Reasoning of the Tribunal

In the original award, the ICSID tribunal found the province had entirely disregarded its own contribution to the water crisis, taking political advantage of it, and that the situation had been caused by years of disinvestment.[99] It stated that the province had more contributed to the crisis than helped to resolve it.[100] In particular, the tribunal held Argentina liable for not having provided "fair and equitable treatment" in light of the legitimate expectations of the investor, as the authorities had acted unreasonably by engaging in political interference related to the tariff regime.[101] Furthermore, the arbitrators held Argentina had breached its obligation to provide "full protection and security", which required, in the arbitrators' view, an overall "stability afforded by a secure investment environment".[102] The tribunal held Argentina also liable for subjecting Azurix to "arbitrary measures" contrary to the BIT by preventing the corporation from collecting payment from its customers.

In two remarkable aspects, however, the tribunal *rejected* Azurix' claims. First, the company had claimed that the amount bid for the concession (the so-called "canon payment" of US$438 million) would justify periodic tariff increases. The tribunal, however, decided that the "canon payment" could not be considered recoverable through periodic tariff increases.[103] It rather fell to the investor to make the appropriate calculation as to the expected earning stream and to bid for the concession accordingly (and not beyond that).[104] This finding became crucial when the tribunal quantified the compensation owed as much lower than Azurix had claimed.[105]

---

[99] ICSID, *Azurix Corporation v. Argentine Republic*, Case No. ARB/01/12, Award of 14 July 2006, par. 144.

[100] *Ibid.*

[101] *Ibid.*, par. 316.

[102] *Ibid.*, par. 408.

[103] *Ibid.*, par. 427.

[104] *Ibid.*, par. 424.

[105] *Ibid.*, par. 442.

Secondly, the tribunal also held that Argentina's actions had not reached the degree of an "expropriation".[106] Argentina had insisted in this regard that the intentions of the State—such as the protection of important public interests such as public health or the right to water—were critical to drawing the line between a "legitimate regulation" and a "confiscatory regulation".[107] The arbitrators expressed support for the approach adopted by an earlier ICSID tribunal in the case of *Tecmed v. Mexico*,[108] where the arbitrators had borrowed a proportionality analysis[109] from the jurisprudence of the European Court of Human Rights in the case of *James v. UK*.[110] If following such an approach, a tribunal should also assess the legitimacy of the aim being pursued by the State with the regulation, in addition to the degree of impact upon the foreign investor. The crucial question would then be whether means and aims were proportionate.[111] The tribunal went on to find that, in this particular case, the degree of impact suffered by the claimant did not rise to the level of an expropriation at all.[112] The tribunal consequently gave no indication as to how it might have assessed the intentions underlying the impugned government actions.

Assessment of the Case in Light of the Human Right to Water

The just described conclusion—not to further explore the issue of expropriation—worked out in Argentina's favor. However, it was, at the same time, most regrettable with regard to the human right to water: it would have been exactly in this weighing—whether the underlying intentions of the State could justify the State's actions affecting the investor—where the right to water could have played a crucial role. Unfortunately, the tribunal did ultimately not have to carry out this weighing of values. At least it indicated its general preparedness to consider public interests underlying the State's action, of which the right to water could then certainly be one. In future cases, the right to water is more likely to be employed to justify a State's action as a legitimate regulation rather than an expropriation, if the right was more clearly accepted.

A second interesting facet in the arbitrators' legal reasoning is the reduced compensation granted to Azurix. The tribunal determined that Azurix was only

[106] *Ibid.*, par. 322.

[107] *Ibid.*, par. 278.

[108] ICSID, *Tecnicas Medioambientales Tecmed S.A. v. The United Mexican States*, Case No. ARB(AF)/00/2, Award of 29 May 2003, par. 121–122.

[109] For the general applicability of such an analysis, Knoll-Tudor (2009).

[110] See ECtHR, *James and others v. The United Kingdom*, Application No. 8793/79, Judgment of 21 February 1986, [1986] ECHR Ser. A No. 98.

[111] ICSID, *Azurix Corporation v. Argentine Republic*, Case No. ARB/01/12, Award of 14 July 2006, par. 311.

[112] *Ibid.*, par. 322.

entitled to "fair market value" compensation for the breaches of the BIT.[113] It even explicitly criticized Azurix for the exaggerated amount it had laid out to obtain the concession as "no well-informed investor would have paid for the concession the price paid by Azurix".[114] Most essentially, the tribunal found that the "canon payment" could *not* be considered as part of the recoverable asset base when it came to setting periodic tariff adjustments. In the arbitrators' eyes, no more than a fraction of the canon payment could realistically have been recuperated under the existing concession agreement.[115] Without naming it, the tribunal hereby applied one of the basic principles of the right to water, as developed in General Comment No. 15: whereas a right to water does not mean a right to free water, it implies a right to affordable water for all.[116] If water services are operated by third parties, the States must prevent these parties from compromising affordable access to water.[117] As a typical violation in respect of the right to water, the Comment names discriminatory or unaffordable increases in the water price.[118] In order to stabilize water prices, States must adopt necessary means, which may include appropriate pricing policies such as free or low-cost water, and water prices must overall be based on the principle of equity.[119] Thus, General Comment No. 15 does not view water as a normal commodity open for profit-making without limits. Water prices must remain connected to the actual costs of providing the service. It was this requirement that ultimately, but invisibly, caused the arbitrators to decide that Azurix was not allowed to fully recover the canon payment through periodic price increases. Implicitly the arbitrators thereby weighed the investors' economic freedom against the people's right to affordable water, and, in their views, the latter outweighed the first.

4.2.2.3 Aguas del Tunari v. Republic of Bolivia[120]

The potentially most prominent investment dispute of the last years related to the privatization of a water supply system is the case *Aguas del Tunari v. Republic of Bolivia*, famously dubbed the "water war" of Cochabamba by the international media, which has been subject to intensive scholarly discussion.[121]

---

[113] *Ibid.*, par. 424.

[114] *Ibid.*, par. 426.

[115] *Ibid.*, par. 429.

[116] ESCR-Committee, *General Comment No. 15*, par. 12 (c) (ii).

[117] *Ibid.*, par. 24.

[118] *Ibid.*, par. 44(a).

[119] *Ibid.*, par. 27, s. 1 and s. 2.

[120] For the facts and the procedural history, see ICSID, *Aguas del Tunari, SA v. Republic of Bolivia*, Case No. ARB/02/3, Decision on Jurisdiction of 21 October 2005, pp. 1–17. Elements of this description and analysis have been drawn from my own previous work, see Thielbörger (2009), pp. 499 ff.

[121] See for instance McFarland Sánchez-Moreno and Higgins (2004); Glennon (2005); Woodhouse (2003); Beisheim (2007); Norris and Metzidakis (2010).

The World Bank had pressured Bolivia in 1997 to privatize the water system of Cochabamba, Bolivia's third biggest city, by making additional aid for water development conditional upon this privatization. In September 1999, after a non-public process with one bidder, Bolivia granted a 40-year concession to take over the water system ("exclusive use" of water sources), previously held by the municipal company Sociedade de Investimento e Gestão (SEMAPA), to Aguas del Tunari, a subsidiary of the American engineering firm Bechtel. By 1997, SEMAPA had only provided water to 57 % of the 600,000 residents of Cochabamba; the system was so inefficient that it lost around 50 % of its water during the transport.[122] Those without access to the system relied on private wells or purchased bottled water.

Within a few weeks after the privatization, prices rose by an average of 35 %,[123] sparking a citywide rebellion. The population even feared that plans to meter wells would cause charges for water that Aguas del Tunari had not even supplied, and that alternative modes of access to water would be reduced or eliminated.[124] In November 1999, massive protests in Cochabamba were violently suppressed by the Bolivian Government. In April 2000, following a declaration of martial law by the Bolivian President, the army killed a protesting Bolivian teenager and wounded more than 100 people.[125] As the citizens of Cochabamba continuously refused to back down, Bechtel eventually gave up the project. Bolivia finally terminated the contract and replaced Aguas del Tunari with a public company again. However, the water supply in Cochabamba is still today considered insufficient.[126] Bechtel subsequently filed a claim against Bolivia under the Netherlands-Bolivia BIT at the ICSID in July 2002. The company claimed US$50 million in damages and profits lost.

Relevant Reasoning of the Tribunal and the Parties

In August 2003, in a "Petition of Non Governmental Organization to Intervene and for other forms of Involvement of Non Disputing Parties", the non-governmental organization (NGO) "Earthjustice" asked permission to take part in the arbitration. Earthjustice applied for party status or status as *amicus curiae*.[127] Additionally, the NGO demanded public disclosure of the statements of both parties and hearings to

---

[122] See statistics at McFarland Sánchez-Moreno and Higgins (2004), p. 1748.

[123] According to some consumers, water prices even increased by 200 %; see McFarland Sánchez-Moreno and Higgins (2004), p. 1763; and Naegele (2004), p. 125.

[124] At the time the unrest began, this had not been the case. However, under the contract, this would have been possible ("exclusive use" of water was granted to Aguas del Tunari); see on this issue Gleick et al. (2002), p. 32; and Naegele (2004), p. 125.

[125] The exact figures differ: see, for instance, Naegele (2004), p. 125; and Glennon (2005), p. 1890.

[126] See Naegele (2004), p. 126; and Glennon (2005), p. 1891.

[127] ICSID, *Aguas del Tunari, SA v. Republic of Bolivia*, Case No. ARB/02/3, Decision on Jurisdiction of 21 October 2005, par. 458.

be held in public. The standing in the dispute derived, so the petitioners, from civil society's "direct interest" in the matter; Earthjustice's involvement would also largely increase transparency.[128]

Already in a letter before any official decision, the president of the tribunal advised the petitioners that only with the consent of both parties, and under the rules set out by the ICSID-Convention and the BIT, could the tribunal allow a non-party to join the proceedings or make any documents public.[129] This consent being absent, the request for public participation was dismissed.[130]

Before the tribunal issued an award, the ICSID procedures were discontinued at the parties' request in January 2006.[131] Bechtel and Bolivia signed an agreement in which they abandoned the case for a token payment of 2 bolivianos (less than one dollar). Both parties stated in their joint declaration that the concession was terminated exclusively because of the civil unrest and the state of emergency in Cochabamba, not because of any act done or not done by the international shareholders. According to Bechtel officials,[132] the company wanted to clarify with this agreement that Aguas del Tunari had acted entirely without fault. In 2007, Bolivia took the decision to withdraw itself from the jurisdiction of the ICSID arbitration.[133]

Assessment of Award and Agreement in Light of the Right to Water

In the *Aguas del Tunari*-case, as in the *Azurix*-case, the missing guarantee of stable prices and the equity of pricing in the concession contract proved fatal. Such an omission is in clear conflict with General Comment No. 15, which demands that services, whether publicly or privately provided, are affordable for all, including socially disadvantaged groups. Equity in this sense means that poorer households cannot be disproportionately affected by water expenses than richer households.[134] Given that the prices increased so dramatically after the privatization in Cochabamba, and poorer households had to spend an enormous part of their budget on water expenses, the right to affordable water was clearly violated.

---

[128] *Ibid.*, par. 458.

[129] *Ibid.*, Decision on Respondent's Objections to Jurisdiction, Appendix III, par. 575.

[130] *Ibid.*, par. 574–576.

[131] See reports in the media, for instance, Harris (2006) or US Water News, Issue February 2006, "Bechtel drops $25 million Bolivian water claim", http://www.uswaternews.com/archives/arcglobal/6bechdrop2.html.

[132] See for instance Statements by Marshall, media relations manager for Bechtel, quoted after Harris (2006). The original statements are no longer available on Bechtel's website.

[133] See, for instance, "Bolivia notifies World Bank of withdrawal from ICSID, pursues BIT revision", *Investment Treaty News*, 9 May 2007, http://www.iisd.org/pdf/2007/itn_may9_2007.pdf.

[134] ESCR-Committee, *General Comment No. 15*, par. 27.

A second important aspect of the *Aguas del Tunari*-case is the jurisdiction of the ICSID arbitration. In Aquas de Aconquija, the tribunal turned down the provisions in the concession contract that had granted exclusive jurisdiction over the contract to Argentina's national courts. In the *Aguas del Tunari*-case, the jurisdiction of the ICSID was based on a BIT which, at first glance, had nothing to do with the State or the origin of the company, namely on the Netherlands-Bolivia BIT. By building up an affiliate and settling it officially in the Netherlands, the US-based company Bechtel launched a clever manoeuvre to bridge the gap that a non-existent BIT of the USA with Bolivia had left open. In this way, it becomes rather easy for companies to create a legal basis for ICSID arbitration, namely by forming an affiliate in a State with whom the host State holds a BIT, and operating the water supply system through this subsidiary. Combining these two findings in Aguas del Tunari and Aquas de Aconquija, ICSID arbitration is almost inevitable after a failed water privatization, regardless of the origin of the company or any jurisdiction-related clauses in the concession contract. In this scenario, the possibility of multinational corporations disrupting democratic processes and evading national laws lies in the air: Art. 24 of the Bolivian constitution,[135] for instance, comparable to provisions in other constitutions, demands all multinational corporations to be subject to national law and jurisdiction. Placing the rulings of international tribunals such as the ICSID above Bolivian law and jurisdiction can thus be considered unconstitutional under Bolivian law. States with similar provisions in their constitutions as Bolivia will consequently consider dropping out from free trade agreements which include dispute settlement mechanisms such as the ICSID, as otherwise they can easily be in breach of their own constitutions. This is a dissatisfying outcome, in particular recalling that the actions against which companies usually proceed before the ICSID arbitrators—even if only in pursuit of their legitimate interests or guaranteed legal rights—are State actions aimed to regulate or protect public interests.

Much relevant in the *Aguas del Tunari*-case is also the cutting down of *existing* access to alternative water sources. This is one of the reasons why the conflict in Cochabamba became so violent: the Bolivian State not only failed to further improve the water supply system of Cochabamba, but granted "exclusive" water rights to the private investor. Thus, existing access to alternative water sources was also impeded—at least according to the wide-spread belief in Bolivian civil society. In doing so, Bolivia not only violated its duty of progressive realization, but also its immediate obligations not to interfere with existing access.

The fourth, and maybe most crucial, lesson to be learnt from *Aguas del Tunari* is the need for and power of public participation. From a formal point of view, no participation by interest groups or civil leaders was possible at the ICSID tribunal:

---

[135] Art. 24 Constitution of the Plurinational State of Bolivia, 1994, http://www.constitution.org/cons/bolivi94.htm, states: "Las empresas y subditos extranjeros estan sometidos a las leyes bolivianas, sin que en ningun caso puedan invocar situacion excepcional ni apelar a reclamaciones diplomaticas".

only the State and the company in question were eligible as parties to the arbitration. However, substantive rights need to be procedurally reflected in order to be effective. It is well possible that interest groups or representatives of civil society would have relied on the right to water when stating their position. However, if the voice is not given the chance to speak, how is the message to be heard? Hence, detention of procedural rights can easily lead to violations of substantial rights.[136] Bolivia should have ensured that the respective BIT allowed participation of third parties in the arbitration process, or otherwise should have made a significant effort to convince the investor to allow third-party participation. In the recent past, arbitrators have started to be more proactive in accepting third party interventions. After third party submissions had been accepted by some tribunals on an individual basis arguing that people could not be excluded if their human rights were potentially implicated,[137] the ICSID amended its arbitrational rules. Tribunals may by now allow written submissions from interested parties, if it contributes a different view than that of the parties and if the non-disputing party has a significant interest in the dispute.[138]

Finally, the right to water was also invoked as a substantial argument by the petitioners in the *Aguas del Tunari*-case. They argued to have a "direct interest" in the matter, which would grant them standing before the ICSID arbitration. Even if in a subtle and elusive way, this is a clear hint to the human right to water. What else could be such a "direct interest", if not the people's right to water? It is exactly the above-mentioned link between substantive law and procedural rights that the petitioners refer to when claiming a "direct interest". It might have been a stronger argument if the petitioners had more explicitly explained their "direct interest" derived from their right to water. However, given the still ambiguous legal status of the right, which I have explained in Chap. 2, it would have invited the arbitrators to turn this reasoning down by simply replying that the right currently is not clearly accepted in international law. Using the vague term of "direct interest" seemed to be less risky for the petitioners in their attempt to get party status; yet it was, at the same time, a missed opportunity for the right to water to play a decisive role in the settlement of an international water dispute.

---

[136] See also McFarland Sánchez-Moreno and Higgins (2004), p. 1777, who argue that the violation of procedural rights might have been the most crucial aspect of the violation of the right to water in ICSID, *Aguas del Tunari, SA v. Republic of Bolivia*, Case No. ARB/02/3, Decision on Jurisdiction of 21 October 2005.

[137] ICSID, *Aguas Provinciales de Santa Fe S.A., Suez, Sociedad General de Aguas de Barcelona S.A. and InterAguas Servicios del Agua S.A. v. Argentine Republic*, Case No. ARB/03/17, Order in response to a petition for participation as *amicus curiae* of 17 March 2006, par. 38–39.

[138] See Arbitration Rules, Rule No. 37(2) (a) and (c), http://icsid.worldbank.org/ICSID/StaticFiles/basicdoc/CRR_English-final.pdf.

4.2.3 The Right to Water as Solution for Challenges of Water Privatization[139]

All three case studies have confirmed what the original assumptions of the preface have already suggested: although meant as a means to improve water supply systems, water privatization can also put human rights related to safe drinking water access at risk. Would a more established human right to water help to resolve this dilemma? Would it make a significant case for the claim that human rights considerations must become more relevant in settling disputes following failed water privatization?

4.2.3.1 Consideration of the Right to Water in the Concession Contract

If a right to water was more explicitly accepted in international law, human rights obligations related to access to safe drinking water are more likely to be considered by the State and the investor when negotiating the concession contract.

Missing or insufficient provisions on the development of future tariffs in concession contracts are often chief causes for later conflicts, as seen in all the three of the examined cases. The question to what extent the price, and possibly a highly exaggerated "canon payment" paid for the concession contract, are recoverable by periodic tariff increases, is often not specified sufficiently in the agreements between States and companies either. Other typical problems are insufficient provisions on quality control (as seen in the *Azurix*-case after the algae outbreak) and the insufficient mandatory investments to be made by investor and State. Equal access to water for everyone is often not guaranteed in the agreements, thereby violating the equity requirement of the right to water as envisaged by General Comment No. 15. Contingency plans in the case of water shortages or dramatic drops in water quality or pressure are often also disregarded by the parties when concluding the agreements. Crucial questions remain unanswered: who is responsible in the case of a water emergency? To what extent are States allowed, or obliged, to interfere in the operation of the private water service? How serious must a crisis be, so that States are allowed to give public warnings or provide emergency water sources?

In an optimal scenario States would simply include stable and clear provisions on these issues in the concession contract. This is, of course, seeing things through rose-tinted glasses. In reality, States are often in a weaker position, whereas investors have the stronger hand. States are under high pressure, due to ailing public finances or to international agencies favoring privatization. The investor,

[139] Elements of this analysis have been drawn from my own previous work in the section "Categories of Violations of the Right to Water and Some Legal and Political Answers", Thielbörger (2009), pp. 503 ff.

to the contrary, understandably only consents to the contract if it appears profitable. This imbalance is even more dramatic where there is only one bidder; States are left without a choice.

However, accepting a human right to water more clearly would raise States' awareness that they deal with a right that is their citizens', not their own, when negotiating concession agreements. They must act as trustees, rather than as the holders of the right. In this capacity, States would have to ensure that the development of prices—and their own possibilities to influence prices—does not take a subordinate role to efficiency rules and economic principles in the contracts. Exaggerated prices, such as canon payments, going beyond all economic reason, lie in the exclusive responsibility of the investor and cannot be recovered through price increases. A purely economic calculation cannot be applied to the question of increasing access to water: the operation of a water supply system is naturally more lucrative in urban than in rural areas. However, it is required under the right to water for everyone to have basic service including humans living in more remote areas. States have to make sure in their agreements that investors do not only focus on optimizing service in lucrative areas, thereby neglecting their duties towards the rural population. Furthermore, States must ensure in agreements that they themselves keep a "safety net function" in cases of emergency. If water quality drops dramatically or if there is an overall water shortage, it is incompatible with the States' protective function for the right to water if clauses in the concession agreement ban the State from warning its citizens, or from providing water from its own water reserves (i.e. those must be kept by States exactly for such for emergency cases.)

A last question is over the period for which privatization can be envisaged. Concessions are given to an investor usually for several decades. This is problematic, especially as the investor, once having signed the contract, inevitably holds the powerful position of a monopolist. These long periods are envisaged for a good reason: investors need time to recoup some of the very costly investments they put into the infrastructure. However, accepting a right to water means that States must address more critically the question whether, and under which conditions, concession contracts granting exclusive water rights over several decades to one investor are compatible with the human right to water. States bear the main responsibility to respect, protect, and fulfill the right. The longer the envisaged time for a concession contract, the more States need to insist on the inclusion of a right to cancel the contract, if necessary. If such a right is not explicitly included in a concession contract, arbitrators will have to include an implied right of cancellation by means of contractual interpretation. Giving water supply out of the States' hand irreversibly can hardly be justified under a human right to water.

4.2.3.2 Requirement of Monitoring and Control

The human right to water would demand that, after water privatization, States have the duty to survey the operation of water assets controlled by the investor and, in cases of emergency, to intervene. Periodic documentation of price and quality development must be undertaken by States. If water prices go up, the water quality drops, or even existing access to water sources is impeded, States have the duty to make immediate use of their possibilities to intervene. These possibilities must be preserved in the concession agreement. The right to water dictates that under no circumstances can States use the concession agreement as an excuse to end their responsibility. The States' responsibility changes after handing over the water supply systems to a private supplier. However, it does not end. To survey whether the operation of the water supplies complies with the right to water, is not an obligation that the States have towards themselves. It is a necessary step to protect the right to water of their people.

4.2.3.3 Inclusion of Procedural Rights

The right to water would also demand compliance with some procedural rights in the case privatization. Procedural rights are necessary for the effective protection of substantive human rights; it is well possible to violate the right to water in its substance by denying the holders of the right their related rights of participation and information. The *Aguas del Tunari*-case has is a prime example for the lack of transparency and public participation: during the time of the negotiation only few opportunities for public input were given and the Bolivian government did not communicate the agreement to the public.[140] Furthermore, the subsequent water law, ratifying the deal between the government and *Aguas del Tunari*, was passed in a hurried manner, which again undermined the possibility for true public participation.[141]

Not ensuring public participation and transparency means violating the right to water itself: it is a logical element of any right to participate in procedures that affect or dispose of the right. In the phase of the arbitrational dispute settlement, it is, oddly enough, typically the ICSID's procedural rules which can cause the procedural violation of the right to water. According to most BITs, participation is only possible if both parties—host State and investor—agree to the third-party participation, or if the tribunal decides to make use of procedural rule 37(2),[142] which at least allows for written submissions. The consent of both parties is not easily reached: investors have an understandable interest in keeping the dispute behind closed doors. In media discussions, they run the risk, rightly or wrongly, of being stigmatized as a profit-seeking "Goliath" in a battle against a "David-State" as the sued country often faces massive economic problems or national debts. Investors understandably want to keep pressure from civil society initiatives out of the arbitrational rooms and out of the arbitrators' heads. It is due to the nature of the ICSID arbitration, which was created to protect investors' rights against States,

[140] For a broader discussion, see McFarland Sánchez-Moreno and Higgins (2004), p. 1747.

[141] *Ibid.*

[142] See discussion about this rule, McFarland Sánchez-Moreno and Higgins (2004).

that civil society and representatives of citizens must be regarded as "third-parties". However, at the same time this appears at odds with the human right to water, as the arbitrators often implicitly decide on the right to water of the people. A more clearly accepted right must therefore demand States to allow public participation to the greatest extent possible. This applies for the time of negotiating the concession agreement, where governments should make public input possible and inform the public about the negotiation and the final agreement as early as possible. It also applies for the time after privatization, when governments must release regular reports about quality control and the development of prices. If it comes to a dispute settlement procedure, States must try to influence the investor to allow third-party participation.

This latter endeavor, however, will often be difficult, if not impossible. Rather than relying on the good will of the investor on a case-by-case basis (when a conflict has already arisen), States must therefore demand the general consent of home States, namely by including provisions on third-party participation in future BITs. More recent BITs, for instance the BIT between Singapore and the USA, already generally allows for third-party participation.[143] This is in line with the US Model BIT from 2004,[144] a prototype document for future BITs between the US and other States. Art. 28(3) of the US Model BIT gives tribunals the authority to accept and consider *amicus curiae* submissions from a person or entity that is not a disputing party. However, it is noteworthy that Art. 28(3) gives the tribunals only the *right* to consider and accept submissions of *amici curiae*; it is in this similar ICSID's recently amended procedural rules. It does not *oblige* tribunals to do so. This is a crucial difference, albeit, as the correspondence in the *Aguas del Tunari*-case between the arbitrators and the NGOs suggests, tribunals might be generally willing to allow third-party participation, but simply have not previously had the legal tools to do so. This would be different under a regime of BITs modeled after the US Model BIT, or if arbitrators regularly made use of Rule 37(2) of the ICSID's procedural rules.

As long as these third-party provisions in BITs are the exception to the rule, the importance of the mobilization of civil society cannot be emphasized enough. It was not the expected outcome of the award in *Aguas del Tunari* that made Bechtel to change course. It was the massive national and international protest in civil society that had started to damage the company's general reputation. On this non-legal battlefield of public discussion and purposeful behavior of customers, the right

---

[143] United States—Singapore Free Trade Agreement, entered into force on 1 January 2004, http://www.ustr.gov/sites/default/files/uploads/agreements/fta/singapore/asset_upload_file708_4036.pdf; see Art. 15.19.3 according to which the tribunal holds "authority to accept and consider amicus curiae submissions".

[144] 2004 Model BIT, Treaty between the government of the United States of America and the government of [country] concerning the encouragement and reciprocal protection of investment, http://www.state.gov/documents/organization/117601.pdf.

to water is an effective argument for NGOs and leaders of civil societies, even before it is more clearly legally accepted. There are no procedural rules that could prevent the invocation of the right to water as a moral right in public discussion.

4.2.3.4 The Right to Water as Mandatory Component of Judicial Review

ICSID arbitration is the single most important institution for the settlement of disputes after failed privatizations, for instance of water supply systems. In accordance with its mandate, it is ICSID's main purpose to protect foreign investment, not to investigate human rights violations. A counterbalance for this ICSID arbitration, namely an explicit global jurisdiction for the protection of the human rights, has been discussed by academics,[145] but is far from becoming legal reality. The recent adoption of the Optional Protocol to the International Covenant on Economic, Social and Cultural Rights (ICESCR) is just a first and small step along this road.[146] However, even if the necessary number of ten ratifications has recently been reached, it remains to be seen how effective the procedure under the Optional Protocol will be.

ICSID arbitrators will remain most crucial in deciding on issues related to the right to water, but, oddly enough, must do so in the absence of the holders of the right (as individuals are not possible parties to ICSID arbitrations), without being able to apply the right to water directly (as the right is not part of the applicable law) and without facing any review of their awards, be it through an appellate review procedure[147] or through decisions of other courts or quasi-courts. The ICSID awards have therefore achieved a status of highest importance in areas that were not originally supposed to be within their jurisdiction.

However, I have pointed out in the three case studies that arbitrators have tried hard to deal with this challenge responsibly. In their struggle to find adequate solutions, they have applied different elements of the right to water, such as affordability or accessibility. However, they had to do so only in implied and unspoken ways. A direct application has never been dared so far. This is regrettable, especially given the principle in customary international law to settle international disputes in conformity with the "principles of justice" in international law, one of

[145] Demanding such a court, see for instance Scheinin (2009); dismissing such a court, Trechsel (2003).

[146] For an overview of the Adoption of the Optional Protocol to the International Covenant on Economic, Social, and Cultural Rights, see, for instance, UNGA, *Third Committee recommends General Assembly adoption of optional protocol to International Convention on Economic, Social and Cultural Rights*, Press Release GA/SHC/3938, 18 November 2008.

[147] See on this the interesting suggestions by Werner (2009).

which should be the universal respect for human rights and fundamental freedoms.[148] However, the arbitrators' reluctance to invoke human rights arguments[149] is owed to the fact that human rights are traditionally not seen as part of the "applicable law". According to Art. 42(1), section 1 of the ICSID-Convention,[150] it is for the parties themselves to choose the applicable substantive law that the arbitrators apply. In the absence of such an agreement, the tribunal will apply the host State's law and international law (Art. 42(1), section 2 of the ICSID-Convention). However, even if the parties do not include international law in their agreement on the applicable law, some tribunals have already stated—from the early 1990s onwards—that they are prepared to ensure the application of certain international minimum standards. The most prominent example in this respect in this regard is the *Southern Pacific Properties v. Egypt*-case,[151] where the parties disagreed about whether the choice of national Egyptian law had been agreed upon or whether international law was applicable in conformity with Art. 42(1), section 2 of the ICSID-Convention. The arbitrators argued that this question was not decisive: they found that, if municipal Egyptian law contained a *lacuna* or if international law was violated by the exclusive application of municipal law, the tribunal would still, in accordance with Art. 42 ICSID-Convention, directly apply the relevant principles and rules of international law instead.[152]

This award is traditionally understood to be favorable of investors, in the sense that a tribunal would not uphold discriminatory or arbitrary actions by a host State against an investor just because these practices are in conformity with the applicable domestic law. However, one can consider the *Southern Pacific* award also in a broader way. This approach to the applicability of principles of international law could also allow the invocation of human rights where appropriate. If the tribunals do not hold back in reading international minimum standards for the benefit of investors into the applicable law, the arbitrators can likewise also uphold minimum standards of human rights law.

The crucial question then, of course, is whether the human right to water belongs to this minimum canon of human rights that under no circumstances can be ignored. It is in this question where the degree of acceptance of the human right to water becomes crucial. Even if the arbitrators were to become more willing in the future to invoke human rights argument in general, will they be willing to apply a right whose acceptance in international law is not beyond doubt?

For the time being, such an assumption remains doubtful. In the meantime, States cannot rely on an interpretation of the applicable law favorable to human rights. They rather have to make sure in the conclusion of BITs and in the privatization

---

[148] See the discussion of these principles of justice and their relevance for investor-State arbitration in Petersmann (2009).

[149] This trend of "reluctance" is also identified by Dupuy (2009) and Hirsch (2009).

[150] ICSID-Convention, 18 March 1965, UNTS 575 (1966), 159.

[151] ICSID, *Southern Pacific Properties (Middle East) Limited v. Arab Republic of Egypt*, Case No. ARB/84/3, Award of 20 May 1992.

[152] *Ibid.*, par. 85.

agreements with investors that the applicable law includes the idea of the right to water. In the absence of such provisions, the ball is in the field of the arbitrators to ensure that minimum human rights standards will always be upheld in their awards. It is in their hands to decide whether they are willing to include the right to water in this canon of rights. An even more explicit recognition of the human right to water would certainly make this decision much easier for arbitrators and other judges.

4.2.4 Conclusion and Remaining Problems

In the settlement of disputes following failed water disputes, the ICSID arbitrators not only rule on the rights of investors, but are implicitly often confronted with essential human rights issues related to water. However, as the three case studies *Aconquija*, *Azurix* and *Aguas del Tunari* have shown, the human right to water in its current status rarely results in an advantage for the States or their people in arbitrational awards and, if so, only in subtle and indirect ways. A clearer acceptance of the human right to water could be a major improvement in two aspects.

Firstly, States would hold a clearer obligation to take special care for the protection and promotion of the right to water of their people long before disputes even arise, namely during all phases of the privatization. Already during the negotiation of their BITs with other countries, States would need to negotiate carefully the applicable law and press for the allowance of third-party participation; during the negotiation of the concession contract, States would need to insist on the guarantee of stable water prices and water quality, and make public input in the negotiation possible; and in the aftermath of the privatization, States would hold a clearer obligation to closely monitor the investor's activity, ensure broad public information and keep the responsibility and capacity for emergency situations

Secondly, if these measures are still insufficient, it is the arbitrators who have to decide whether they nevertheless want to pay tribute to the right to water in their awards. The arbitrators have different options in how to do so. Firstly, they can invoke the human right to water as one of the "principles of justice", thereby using it as a tool for treaty and contract interpretation by readings its principles into the BIT and the concession contract. Secondly, and more promisingly, tribunals can consider the right to water as part of the applicable law, in understanding this term as part of a minimum human right standard that cannot be neglected regardless of what the parties agreed upon as the "applicable law". An overall clearer recognition of the human right to water as a legally binding human right would thus be of great help for the investment arbitrators: their double dilemma[153] would then have become a single one. Arbitrators would only have to decide whether they regard human rights as part of the "principles of justice" or the "applicable law". They would no longer have to question whether the right to water is one of these human rights.

[153] I have coined this phrase previously in Thielbörger (2009), pp. 487 ff.

4.3 Realization by International Means: The Case of International Obligations Deriving from a Right to Water

Chapter 4 has so far found that effective international monitoring mechanisms for the right to water are in place. Secondly, it is likely that with increasing recognition of the right it will also be more and more enforced against the private sector, in particular in arbitration procedures settling water disputes between States and investors. A third crucial element of implementing the right to water would that of not only realizing it through national means, but through international efforts.

Traditionally, the realization of human rights obligations has been understood as limited by territory and jurisdiction.[154] However, developing countries often cannot master their water crises or ensure the provision of economic and social rights for their people on their own. International assistance and cooperation is needed. This chapter explores the legal and philosophical dimensions of such international water assistance and cooperation. It focuses on the obligation of particular donor States and the community of States as a whole. It retraces the path of international obligations with regard to the right to water: from where can they legally and philosophically emerge, and where in international law and politics are they concretely headed?

4.3.1 International Water Obligations: The Philosophical Background

Chapter 4 deals with the question of how a right to water could be made effective in practice. However, in addressing the doctrinal evolution of international obligations for the right to water, it is of primary importance to also determine plausible theoretical bases for such obligations: why should we even strive for practical ways of international assistance, if morally there is no compelling reason for it?

The most critical component of such international assistance is that of "*fulfilment*". Thus, the question is whether an international "negative" failure to assist in the realization of the right can be seen as an equal harm as the harm done through "positive" action. Is there sound cause to assume that States are responsible for harms outside their territory that are not directly or causally attributable to their own misdoings? The classic formulation enumerated in the Articles on State Responsibility[155] would only go as far as recognizing a duty to refrain from causing transboundary harm. In recent decades, however, one has witnessed a move away from

[154] See e.g. Art. 2(1) ICCPR (1966), UNTS 999 (1976), 171; Convention against Torture and Other Cruel, Inhuman or Degrading Treatment or Punishment, 10 December 1984, UNTS 1465 (1987), 85.

[155] UNGA, *Articles on the Responsibility of States for Internationally Wrongful Acts*, A/RES/56/83, 28 January 2002, Annex.

consent-based sovereign conceptions to community or people-centred approaches to international law. But how does one determine the moral or political responsibility of States for people outside their territory and jurisdiction?

4.3.1.1 A Liability Model Approach to the International Obligation to Fulfill

One possible approach is to equate responsibility with fault. This might be termed a *liability model* of responsibility.[156] This conception is reflected in the well-established "polluter pays" principle[157] in international environmental law, as mentioned for instance in Principle 16 of the Rio Declaration on Environment and Development.[158] Responsibility is assigned on the assumption of a clear and identifiable agent.

This approach, however, is entirely at odds with an international obligation to *fulfil* human rights. It is almost impossible to identify the responsible agents for most water crises. In cases of natural disaster, there often even are no responsible actors at all. Following an approach along these lines, would essentially mean not to assume any obligation to fulfil; an assumption that is at odds with the by now widely accepted triad of human rights obligations.[159]

4.3.1.2 Some Kantian and Consequentialist Approaches

If we look at moral responsibility to fulfil from a more classical philosophical perspective, we might consider some de-ontological and consequentialist arguments.[160] Unsurprisingly, both Kantians and consequentialists would agree that it is bad for humans to die of starvation and thirst. Where both philosophical streams would tend to disagree, however, is, in how this situation affects the moral obligations of those who watch this starvation from afar. The main distinction is that Kantians would argue that the principal moral obligation of developed States is

[156] See for a good (but in conclusion dismissive) assessment of this topic for instance Young (2004), in particular pp. 366–367.

[157] See on this topic for instance Singer (2004), pp. 27 ff.; Boyle (2009); Mamlyuk (2009), pp. 40 ff.

[158] UNGA, *Report of the United Nations Conference on Environment and Development: Rio Declaration on Environment and Development*, A/CONF.151/26 (Vol. I), 12 August 1992, Annex I, Art. 16: "National authorities should endeavour to promote the internalization of environmental costs and the use of economic instruments, taking into account the approach that the polluter should, in principle, bear the cost of pollution, with due regard to the public interest and without distorting international trade and investment.", http://sedac.ciesin.org/entri/texts/rio.declaration.1992.html.

[159] OHCHR (2005), pp. 15 ff.; Leckie and Gallagher (2006), pp. xx ff.; Riedel (2011), par. 14–15.

[160] See as examples for consequentialist approaches Shaw (2006); and, more specifically on the question of famine and thirst, Singer (1972). As an example for modern Kantian approaches, see Hill (2006), and, more specifically on the question of famine and thirst, O'Neill (1980).

to ensure that they do not treat the starving as mere means; consequentialists, like Peter Singer, would, to the contrary, construct a strong duty in favour of active intervention to stop the starvation of people in developing countries.

One of the Kantian key arguments would be the distinction between positive and negative obligations. Kantians would argue that we are under a *negative* duty to ensure that we do not treat the starving as mere means, particularly not in order to exploit their weaker economic position. At the same time, we are not necessarily obliged to *positively* alleviate their poverty.[161] Although we *ought* to provide help to others, we do not owe it to them. In this sense, there is no such thing as an international "duty of justice" in the Kantian approach: any international help would be charity rather than justice.[162] Consequentialists would, of course, counter that this approach fails to recognize that the dichotomy between negative and positive duties is often a false one: we are not only responsible for what we do, but also for what we fail to do, as the consequences of our action or inaction are often equivalent.

Some of the Kantian theorists have tried to move away from this classical strict dichotomy between positive and negative duties and to create obligations that go beyond Kants' minimalist conception of "duties of justice". Onora O'Neill for instance emphasizes the idea of complementing "duties of beneficence".[163] These require acts that promote others' ends and that foster others' capacities to pursue *their own ends* as autonomous beings. These "duties of beneficence" become especially important for those parts of the world where extreme poverty and hunger leave people unable to pursue any of their own ends. As such, the relief of hunger and thirst, according to O'Neill, must rank very high among duties of beneficence.[164] It is a duty that does not paternally urge us to come to the aid of others, but rather ensures that others are equipped with the means and tools to realize their goals as autonomous beings.

4.3.1.3 Rawlsian-Influenced Approaches

Contemporary thinkers in the field of political philosophy have keenly observed a transnational dimension to political and legal obligations. For instance, Thomas Pogge[165] has criticized models of responsibility that are purely based on fault or liability, arguing instead that political responsibility extends beyond the boundaries of the State and their populations. Pogge asserts that some obligations of justice are rather owed to global recipients.[166] This challenges the orthodox belief that States

---

[161] Cranston (1973), p. 68.

[162] Buchanan (2004), pp. 195 ff.

[163] O'Neill (1980), p. 292.

[164] *Ibid.*

[165] See for instance Pogge's most influential book, *World Poverty and Human Rights*, (Pogge (2002)), therein: chapter 8: Eradicating Systemic Poverty: Brief for a Global Resources Dividend.

[166] *Ibid.*, p. 203.

are responsible only to their own constituencies. What is in particular relevant to our analysis here about Pogge's arguments about the eradication of world poverty is his attempt to rely on a *negative* rather than a *positive* duty.[167] According to Pogge, the global economic elite has a stringent duty of justice to progressively eradicate global poverty, because they have infringed upon their *negative* duty not to contribute to the establishment of a global institutional order that, avoidably and foreseeably, leaves the basic economic and social rights of others unfulfilled.[168] Pogge states that "[t]his does not mean that we should hold ourselves responsible for the remoter effects of our economic decisions. [. . .] But we must be concerned with how the rules structuring international interactions foreseeably affect the incidence of extreme poverty."[169] Thus, Pogge's conception of global political responsibility broadens the horizon of political responsibility to include the network of structures and institutions that bring about the conditions of deprivation and degradation for certain affected populations.

Pogge's argument is very much in line with the *Rawlsian* idea of "justice as fairness"[170] developed through John Rawls' landmark "A Theory of Justice".[171] According to Rawls, each person has an equal right to basic liberty to the extent that the exercise of that liberty is compatible with a similar liberty for others. Social and economic inequalities have to satisfy two conditions: they must be attached to positions open to all under fair and equal conditions ("liberty principle") and these inequalities have to be to the benefit to the least advantaged members of society ("difference principle").[172] The difference principle thus permits inequalities in the distribution of goods *only if* those inequalities benefit the worst-off members of society. Each member of society has an equal claim on their society's goods so that inequality is acceptable only if it is to the advantage of those who are worst-off. If we translate this "difference principle" from a domestic to a cosmopolitan level, combining it with the approach of Pogge, this would create a moral obligation for the "better-off" to promote the benefit of the "least advantaged", as it is the global legal order that has given them this advantage, e.g. by helping to fulfil human rights obligations in their respective countries, but not in others.

4.3.1.4 A Water-Specific Approach to International Water Assistance: Hugo Grotius

This general approach to global justice must, however, be applied to the more specific issue of water. How, if at all, could we justify the claim that the water-rich

[167] *Ibid.*, pp. 203 ff.

[168] *Ibid.*

[169] *Ibid.*, p. 206.

[170] Rawls (1958).

[171] Rawls (1971).

[172] *Ibid.*, pp. 60 ff.

world has to share its water with water-poor regions? Do the water resources of the world belong to all people or do a State's water resources attach solely to its territory alone?

Hugo Grotius and the Free Sea

We might start with the proposition that water is a common resource of all humankind. The idea that water must receive special philosophical consideration, and cannot be possessed by any State, is not new. The Dutch philosopher Hugo Grotius already wrote in the seventeenth century his dissertation about "The Free Sea".[173] Grotius' main idea was that the sea could not be owned by anyone, and that no nation could have exclusive rights, such as passage or trading rights, over it. The sea should not be treated like land (which could be owned by nations), but more like air: the sea should be considered as the common property of all.

Grotius asserts two reasons for this argument. He claims firstly that the sea is not "susceptible of occupation", and, more importantly for the purposes of our analysis, that secondly it is "destined for the use of all".[174] For his second argument, he refers to Roman philosophers and also gives religious affirmations.[175] Grotius calls things like air and water common to all men "by natural law."[176] Although a heavily contested idea at Grotius' time,[177] the modern law of the sea follows Grotius' general idea of the freedom of the high seas.[178] Different from several zones of limited usage rights along the coast,[179] the high seas cannot belong to any State. It can be used by anyone for peaceful purposes.

Applicability of Grotius' Theses on International Water Assistance

However, the applicability of Grotius' theory onto international water assistance is, of course, limited. It holds three main caveats. Firstly, Grotius, by speaking about seawater, was making claims about (non-drinkable) saltwater, not about freshwater. The subject of water assistance, however, is freshwater. Secondly, Grotius was clearly focussing on water bodies rather than transferable water. And thirdly, Grotius was referring to those water bodies not enclosed in one State's territory. He was mainly dealing with the high seas, not with inland water sources.

---

173 Grotius (2009).

174 *Ibid.*, p. 28.

175 *Ibid.*, referring explicitly to Cicero and Ovid, pp. 27–28.

176 *Ibid.*, p. 29.

177 See for instance Welwood (1613); *id.* (1615); as well as Selden (1636).

178 See Art. 87(1) UNCLOS, 10 December 1982, UNTS 1833 (1994), 3.

179 See Art. 2 ff. UNCLOS on the territorial sea; Art. 33 UNCLOS on the contiguous zone; and Art. 55 ff. UNCLOS on the exclusive economic zone.

Keeping these limitations in mind, Grotius' arguments are still relevant for the issue of international water assistance today. Especially Grotius' second reason—that waters are destined for the use of all—is worth considering for the case of international water assistance. If this reason could apply also to inland freshwaters, and thereby to the main sources of drinkable water, one can build a similar moral reasoning with regard to drinking water as Grotius did with regard to the high seas. Grotius claims that the seas are "destined for the common use of all".[180] According to Grotius, some things cannot be private possession, but they are "by nature things that are open to the use of all, because in the first place they were produced by nature", thus, "they have been created by nature for common use".[181] As they belong to human society as a whole,[182] it is ultimately the consensus of opinion of humankind that exempts things such water from private ownership.[183] Thus, Grotius' argument is that both land and sea were originally common property. Most land became private property because, in being used by individuals, it became less fit for the use of others. However, this does not apply to the sea: Grotius argues that, according to the law of nature, an individual must allow another the use of something if the use does not result in loss to himself. Understanding Grotius' argument in this way, it becomes applicable also to freshwater: water is generally a renewable resource. Its use does not mean a loss to anyone in the long run, as long as there is no water over-use and the ability to regenerate is not damaged.[184] Following Grotius, water-abundant States—whose water is plentiful and renewable—would then have a moral obligation to allow the use of water for water-scarce countries.

4.3.1.5 The Idea of Common Ownership of the Earth

Contemporary philosophers have used Grotius' idea to go even further and argue for the *common ownership of the earth*. Building on John Locke's idea of the common ownership of the earth,[185] Mathias Risse has recently argued that humanity collectively owns the resources of the earth.[186]

These arguments mainly aim at the area of habitable lands and natural resources, being unevenly distributed amongst different nations and people. As a consequence, so Risse argues, one must allow for immigration: big countries with low population density and many resources must allow access for immigrants from

[180] Grotius (2009), p. 28.

[181] *Ibid.*

[182] *Ibid.*

[183] *Ibid.*, p. 29.

[184] See the Argument of Hardin on the "Tragedy of the Commons".

[185] Locke (1821), chapter V, section 26, sentence 1 ("God, who hath given the world to men in common [. . .]").

[186] Risse (2008a); *id.* (2008b); *id.* (2012).

countries with high population, little land and few vital resources. However, the converse argument can be made for the case of water assistance: if humanity owns the grounds of the land as a whole, with all its consequences for the moral permissibility of immigration, it is manifest that also the freshwater sources of the earth belong to humanity as a whole in the form of "common ownership".[187] One possible consequence would be to allow for water-migration: one must allow people to settle near watercourses and in water-abundant countries. Another, and more feasible, consequence is mandatory water assistance: as water is more flexible than solid land, water-abundant countries could simply have to share their fresh water with people in water-scarce countries in various forms of water assistance.

4.3.1.6 Conclusion: A Compelling Philosophical Case for International Water Assistance?

Traditional views that rely on Kantian ethics or on a "liability as responsibility model" have problems to recognize an international obligation to fulfil human rights such as the right to water. Especially the liability model is inadequate for the example of a right to water where it is difficult to identify who is at fault for water shortages in the first place.

A consequentialist approach, however, would be more inclined to accept international duties: because of a moral duty to maximize happiness, regardless of country or origin. More progressive Kantian views would agree with this result by accepting an international duty on the grounds that law must provide the conditions by which individuals can become fully autonomous beings.

Modern philosophers, such as Pogge, have emphasized the need for an approach that includes ideas of global justice and global fairness: the wealthy part of the world cannot remain effortless, if they have helped to create and reinforce a global institutional framework that has created this global injustice. If society is a *global* society, any increases in global wealth must not only benefit a rich elite of nations, but also benefit, in a Rawlsian sense, the least well-off too.

Approaches that rely on the specific case of water come to no other conclusion. A morality based on the ideas of Hugo Grotius' views on the free see, would argue that also freshwater, like the waters of the High Seas, was destined for the use of all people. Water-abundant countries must share their water, as long as this does not lead to a loss of that water in the long run. Modern theories on global justice and the of common ownership of the earth, like Risse, would even go one step further: either we must allow the people to move to places where fresh water is abundant ("water immigration") or we must move the water to the people that need it

[187] So also Risse in his newest paper, Risse (2013). Risse bases his assumption again on Locke (1821), in particular chapter V, section 29 "the water in the fountain be everyone's" and section 33 "the case of land and water, where there is enough of it, is perfectly the same".

("international water assistance"). If confronted with this choice, water-abundant countries might prefer the second option: it is more practicable and less costly for water-abundant States.

Traditionalists will certainly contest this view and favour the traditional approach that focuses on the State as the only duty bearer within its territory and jurisdiction. However, it is noteworthy that such an approach is no longer the dominant view, certainly not in political philosophy. Many approaches in contemporary moral philosophy concerning issues of global hunger and thirst, or common ownership of the earth, offer convincing reasons of how to justify an obligation to fulfil human rights also through international efforts. If water was accepted more clearly as part of these human rights, a compelling moral case for mandatory international water assistance can certainly be made.

4.3.2 International Water Assistance: Existing Legal Implications

Which obligations would an international water assistance based on the acceptance of a human right to water entail?[188] The answer is twofold: On the one hand, some international treaty provisions on human rights refer already to transnational and international obligations related to international assistance and cooperation. On the other hand, with the increasing acceptance of the right to water, some of these obligations would be made more explicit, and some more new obligations might even be added.

One should start by distinguishing the terms international *assistance* and international *cooperation*. The US have suggested to use both terms interchangeably during the drafting process of the ICESCR.[189] Recent treaties tend to prefer the term "international cooperation" over that one of "international assistance".[190] In order to avoid terminological fuzziness, this chapter will follow the definition suggested, *inter alia*, by Matthew Craven, who defines cooperation as all forms of "mutual action directed towards a common goal" whereas assistance is understood as "the provision or transfer of some "good" from one State to another".[191]

---

[188] I have given a more detailed account on the relationship of the human right to water and international water obligations—albeit with a specific focus on the Euphrates Tigris region—in: Thielbörger (2013), chapter 3.

[189] The US representative stated that "the words 'international co-operation' in the original text of article 2, paragraph 1 adequately covered all forms of international assistance", quoted after Skogly (2006), p. 87 citing UNGA, Seventeenth Session, Third Committee 1204th meeting, Official Records, par. 49.

[190] See for instance Art. 4, 17 and 24(4) CRC, 20 November 1989, 1577 UNTS, 3; or Art. 4 and especially Art. 32 UNGA, *Convention on the Rights of Persons with Disabilities*, A/RES/61/106, 13 December 2006.

[191] Craven (1995), pp. 146–147.

4.3.2.1 International Obligations in Human Rights Instruments

As a general rule in international law, human rights obligations are limited to a State's territory and jurisdiction; this is in particular recognized for civil and political rights.[192]

However, there are some notable references to international obligations in the norms of international law as well. Art. 1(3) of the UN Charter names as one purpose of the UN to achieve international co-operation in solving international problems of economic, social, cultural or humanitarian character, thereby promoting and encouraging respect for human rights and fundamental freedoms. Similarly, Art. 56 of the UN Charter obliges all members to commit themselves to take joint and separate action for the achievement of the purposes set forth in Art. 55 of the Charter, such as economic, social or health-related goals. Furthermore, the Universal Declaration on Human Rights (UDHR) recognizes a variety of international obligations.[193] Art. 22 of the UDHR states that everyone is entitled to the realization of their economic, social and cultural rights "through national effort and international cooperation [. . .]". Art. 28 of the UN Declaration even entitles individuals to a social and international order in which the rights of the Declaration can be fully realized. These examples emphasize that, in spite of human rights being generally tied to territory and jurisdiction, the drafters of both the UN Charter and of thc UDHR were somewhat aware that the goals of the UN, in particular the promotion of human rights, are only possible through international collaboration.

The provisions of the ICESCR make this case even more explicitly. According to Art. 2(1) ICESCR Member States must take steps individually and *through international assistance and co-operation*, especially economic and technical one, to achieve progressively the full realization of the rights of ICESCR. The right to an adequate standard of living (Art. 11 ICESCR), from which—as I have argued in Chap. 2—parts of the right to water is derived, recognizes the essential importance of *international cooperation* based on free consent. Art. 11(2) ICESCR furthermore recognizes the individual right to be free from hunger; States must take the necessary measures, individually and through *international cooperation*, in order to ensure, *inter alia*, an equitable distribution of world food supplies in relation to respective need.

4.3.2.2 Assessment of Existing International Water Obligations

Economic and social rights have a stronger focus on international obligations than civil and political rights. Although the drafters of treaties with respect to socio-

---

[192] Art. 2(1) ICCPR (1966), UNTS 999 (1976), 171.

[193] As was argued earlier in this book, UNGA, *Universal Declaration of Human Rights*, Res. 217 A (III), A/810 at 71, 10 December 1948, is technically not legally binding; however, it is in large parts considered part of customary law.

economic rights were aware that these rights needed stronger international cooperation and international assistance to be realized, none of the relevant documents offers a definition of the terms of assistance or cooperation. States are divided whether such obligations can, in fact, exist. Unsurprisingly, rich donor countries tend to deny the existence of a legal obligation to provide international assistance,[194] whereas developing countries, including many of the Group 77 States, insist that international assistance and cooperation are legal obligations.[195] Academics have also partly critiqued this denial of responsibility and insisted on such a legal obligation, following either directly from the above examined treaties or from an emerging norm of international customary law.[196]

4.3.3 International Water Obligations Under an Accepted Right to Water

If we accept that human rights, in particular socio-economic rights, can entail some international dimensions, what international obligations can be identified under a human right to water? General Comment No. 15, albeit not a legally binding document, makes reasonable and concrete suggestions and dedicates a whole chapter to international obligations.[197] Similar to the general triad of State obligations with regard to human rights, international obligations can also be subdivided into duties to respect, protect and fulfil.

4.3.3.1 International Obligations to *Respect* the Human Right to Water

The obligation to respect applied to the right to water generally means to not interrupt an existing water supply.[198] There are different scenarios imaginable in

[194] See for instance statements of Canada and France in ESCR-Committee, *Report of the open-ended working group to consider options regarding the elaboration of an optional protocol to the International Covenant on Economic, Social and Cultural Rights on its second session*, E/CN.4/2005/52, 10 February 2005 (hereinafter "ESCR-Committee, *Report on an optional protocol to the ICESCR*"), par. 76 ff. On a similar debate concerning the right to development, see HRC, *Report of the Working Group on the Right to Development on its eighth session*, A/HRC/4/47, 14 March 2007, par. 52 and Annex III, par. 63.

[195] See for instance statements of Nigeria and Egypt in ESCR-Committee, *Report on an optional protocol to the ICESCR*, par. 76. In the debate concerning the right to development, see the statement of Algeria on behalf of the African States, in the Report of the Working Group on the Right to Development, par. 18, and of Cuba on behalf of the Non-Aligned Movement (NAM), par. 19 and Annex III, par 62.

[196] Alston (2005), p. 778; Sepúlveda (2006); Baderin and Ssenyonjo (2010), pp. 67 ff.; Bueno de Mesquita et al. (2010), pp. 112 ff.

[197] ESCR-Committee, *General Comment No. 15*, par. 30–36.

[198] *Ibid.*, par. 10.

which a State other than the host State carries the obligation to respect the right to water. A State might for instance contaminate shared waters such as rivers or lakes, thereby negatively affecting the access to clean water of the people of another (e.g. neighbouring) State.[199] In international environmental law, the famous *Trail Smelter*-case[200] has already emphasized in the early 1940s that a State must not cause damage to the environment of other States. The tribunal in that case argued that in international law no State has the right to use or permit the use of its territory in a way that could cause injury by fumes to the territory of other States or their populations.[201] This generally-accepted principle of environmental law has also notably been repeated in Principle 21 of the Stockholm Declaration of 1972 and in Art. 7 of the Watercourse Convention of 1997.[202] This principle, although originally evolved in environmental law, must equally apply to the protection of human rights, and to the right to water, if accepted as one of them. While the *Trail Smelter*-case established rights between States (as the traditional subject of international law) rather than accepting the rights of people, polluting a water supply system of another country will always be both, an interference in the domestic affairs of that State and an infringement of the right to water of its people. In that sense, it is increasingly recognized in international law that the obligation to respect socio-economic rights cannot stop at a State's borders, but is required from a State concerning the populations of all countries.[203]

Turning to the right to water specifically, General Comment No. 15 suggests how such international obligations could translate into obligations under the right to

[199] Note in this respect also the *Convention on the Law of the Non-navigational Uses of International Watercourse*, UNGA, A/RES/51/229, 8 July 1997, Annex. The Convention proscribes in Art. 7 an obligation not to cause significant harm to other watercourse States and in Art. 5 and 6 the principle of equitable use. However, although drafted already in the 1990s, the Watercourse Convention has still not entered into force: only 30 (as of 24 February 2013) States have ratified the Convention; 35 ratifications would be necessary (Art. 36(1) Watercourse Convention). For the current status of ratifications, see the UN's official website at http://treaties.un.org/Pages/ViewDetails.aspx?src=UNTSONLINE&tabid=2&mtdsg_no=XXVII-12&chapter=27&lang=en#Participants.

[200] *Trail Smelter Arbitration* (United States v. Canada), Award of the US-Canadian Arbitral Tribunal of 16 April 1938 and 11 March 1941, UNRIAA Vol. III (1949), 1905.

[201] *Ibid.*, see in particular 1965.

[202] United Nations Conference on the Human Environment, *Declaration on the United Nations Conference on the Human Environment*, A/CONF.48/14/Rev.1, at 3, 1973, adopted on 16 June 1972, Principle 21: "States have [. . .] the sovereign right to exploit their own resources [. . .] and the responsibility to ensure that activities within their jurisdiction or control do not cause damage to the environment of other States or of areas beyond the limits of national jurisdiction."; and Watercourse Convention, 1997, Art. 7: "Watercourse States shall, in utilizing an international watercourse in their territories, take all appropriate measures to prevent the causing of significant harm to other watercourse States.".

[203] For instance, Committee on the Rights of Child, *Day of General Discussion on "Resources for the Rights of the Child—Responsibility of States"*, 5 October 2007, http://www2.ohchr.org/english/bodies/crc/docs/discussion/recommendations2007.doc, par. 51.

water. Par. 31 of General Comment No. 15 contains the general obligation to respect the enjoyment of the right to water also in other countries, and obliges States to refrain from interfering with the right in other countries. Embargoes or any other measures that would prevent the supply of water in another country are specifically banned as instruments of political or economic pressure (par. 32). States must for instance refrain from any action that would negatively impact an existing water access within another State's borders when they are given the opportunity to do so (e.g. during times of an international mission in that State). If international humanitarian water assistance is provided to a State within times of crisis (e.g. following a natural disaster), this must be done in a non-discriminatory manner, so that the whole population, and not only one class or one ethnic group benefits: the duty of non-discrimination under the right to water is an immediate one, independent of available resources.[204] The international dimension of the obligation to respect holds also special relevance for the relationship of upstream- and downstream States: upstream States must not decrease or supress the amounts of water they release (e.g. through dams) in order to put political pressure on their downstream neighbours.

4.3.3.2 International Duties to *Protect* the Human Right to Water

Traditionally it is the State who has to prevent third parties from interfering with rights such as the right to water. In particular situations, however, this duty to protect might have an international dimension. Par. 33 of General Comment No. 15 suggests that States, or the community of States, must influence third parties through political and legal means in accordance with the UN Charter and the applicable international law to respect the right to water also in other countries.[205] This case becomes relevant mainly in the context of the privatization of water supply systems through foreign investors. If these water privatizations fail, not only the host State, but also the home State of the investor might hold a special duty to protect the right to water. In fact, it is very often the case that the private investor has its headquarters in a State other than the host State.[206] The task for these States then is to regulate the transnational water companies headquartered in their jurisdiction, if legally possible, and prevent them from doing harm to the right to water in other countries. Similarly, if certain uses of shared water bodies are managed

[204] ESCR-Committee, *General Comment No. 15*, par. 17.

[205] This is a very common description of the right to protect as an international responsibility. ESCR-Committee, *General Comment No. 14* on the right to health, for instance, holds a very similar description, see par. 39.

[206] Many water companies are in particular headquartered in France and the UK, see Hall and Lobina (2007), pp. 68 ff. Also see the analysis undertaken in Sect. 4.2.2 of this book.

by joint bodies,[207] the States which are legally in a capacity to do so must regulate these joint bodies in their domestic law, thereby not only protecting the right to water in their own, but also in the neighbouring State against illegal interference.

4.3.3.3 International Obligations to *Fulfill* the Human Right to Water

The most contested category of international obligations, and at the same time the "heart" of the debate, is that of an international obligation to *fulfil*. Given the potential far-reaching consequences of such international obligations to fulfil, one must be particularly careful with assuming such obligations all too easily.

However, certain categories of such international duties to fulfil are identifiable. One example is that of former colonialism. Where colonial exploitation has done lasting damage to another State's water resources or its ability to provide water services to its people, thereby making the fulfilment of the right in that State difficult, this can create special responsibilities for the former colonizer. A second example for a specific international obligation to fulfil can arise from the voluntary acceptance of such a responsibility. In the case of close economic or political unions, as for instance the EU,[208] members of such constantly intensifying unions have started to take over intensified common responsibilities. These duties often exceed the bare duty not to do harm to the other members' economic or ecological systems, but rather entail the duty to support the fulfilment of socio-economic rights in the other member States. How these duties are shaped *in concreto*, is a question of the law agreed between the different Member States. Being (partly) responsible for an economic crisis in other States, which in turn might make the development and improvement of water supply systems difficult, is, however, certainly not enough to create a particular obligation. Otherwise global economic crises like the current one would create grotesque responsibilities for the creators of that crisis, which are themselves often already significantly affected by the crisis itself. Similarly, neither the bare fact of a former colonial relationship between two States nor the bare common membership in a joint economic or political union can be considered sufficient to trigger such relationship. The question of whether a specific international obligations exists or not must be decided on a case-by-case basis. However, it is noteworthy that, in particular cases, there are good reasons to attach the duty to fulfil the right to water rather to several specific States than to that one State alone in whose territory the fulfilment of the right to water is not sufficiently developed. Typically, however, the international dimension of the duty to *fulfil* the right to water would be with the community of States as a whole duty-bearers, not

[207] See for instance the Itaipu Dam Project (a dam project on the Panama river between Brazil and Paraguay), operated by Itaipu Binacional, http://www.itaipu.gov.br; or the Lesotho Highlands Water Project (a joint water project between South Africa and Lesotho) operated by a joint commission, htttp://www.lhda.org.ls.

[208] See the analysis in Sect. 2.1.2 of this book.

with specific States. The duty will fall upon all those States that are in a position to provide such international water assistance. Not all States may be in a position to provide *financial* international assistance. However, a large number of States might be capable of providing technical water assistance or of concluding international water treaties; treaties that allow for the fulfilment of the right to water in the contracting States. The notion "in a position to do so" here is of crucial importance. It should be understood in a similar way to the more established principle of "common but differentiated responsibilities", as recognized as a soft law principle in international environmental law[209]: the depletion and unequal distribution of water, the argument would go, imposes special duties upon developed or water-abundant countries to assist in the fulfilment of the economic and social duties attached to the right to water in water-scarce or generally less developed States.

General Comment No. 15 addresses the obligation to fulfil in par. 34, albeit without naming it explicitly. It suggests that the duty consists of two main elements: the obligation to *facilitate* and the obligation to *provide*. Depending on the availability of resources, to "facilitate" means an obligation of creating an environment in which the population of other States is able to increasingly enjoy their right to water.[210] This obligation might include taking the initiative for exchange of expertise in the field of water and supporting further research on water-management in that respective State. In terms of the obligation to "provide" international water assistance, General Comment No. 15 addresses this issue by suggesting that it is in particular incumbent on State parties to provide international assistance and cooperation, especially economic and technical, to enable developing countries to fulfil their core obligations.[211] It is noteworthy that the General Comment in par. 38 finally makes use of the term obligation to "fulfil". However, reading carefully, par. 38 refers to "their" (ergo: the developing countries') duty to fulfil. It falls on developed countries to *help* in fulfilling this duty, thereby indirectly suggesting that the developed States do not have a duty of their own. Par. 35 of the General Comment also calls upon State parties to give due attention to the right in international agreements and recommends to consider the development of further legal instruments on the right to water. In international and regional agreements in the future, States should thus take steps to ensure that the right to water is not adversely affected. In particular, agreements on trade liberalization and water privatization should not affect countries' capacity to fulfil the right to water: an interesting provision certainly in particular for the conclusion of future BITs.[212]

---

[209] See on this principle for instance Harris (1999); Ali (2002); Hey (2011).

[210] Compare ESCR-Committee, *General Comment No. 15*, par. 25.

[211] *Ibid.*, par. 38: "For the avoidance of any doubt, the Committee wishes to emphasize that it is particularly incumbent on States parties, and other actors in a position to assist, to provide international assistance and cooperation, especially economic and technical which enables developing countries to fulfill their core obligations indicated in paragraph 37 above."

[212] See discussion of States' duties in this respect, at Sects. 4.2.3.1 and 4.2.3.3.

Finally, Art. 36 calls upon States, in their capacity as members of international organizations and international financial institutions (IFIs), to develop lending policies, credit agreements and other international measures that are consistent with the right to water.

Altogether, General Comment No. 15 suggests that the international obligation to fulfil is the exception, not the rule. It should be understood to focus on the realization of core obligations. Establishing some international obligations to fulfil is not the general attempt to equalize water conditions in developing and developed countries altogether. It is rather to ensure that the basic standard of the fulfilment of the right to water can be met everywhere.

4.4 Conclusion and Remaining Problems

A clearer acceptance of a human right to water in international law would make international obligations with respect to water much more explicit and raise awareness for international water issues beyond the general concept of international development. The right to development[213] is in fact a deterrent example for the right to water: a right that never dared to find its contours and always remained ineffective. In order for a right to water to be effective, it will need to fledge out those obligations that fall upon the international community, and those, that do not. General Comment No. 15—despite its non-legally binding character an excellent guide to possible international obligations—makes a remarkable effort to identify positive and negative international obligations with respect to water: it suggests international steps to be taken in order to advance the full realization of the right, while also identifying those international measures that under all circumstances are inconsistent with the right (e.g. water embargoes as a means of exerting political pressure). The General Comment finds formulations that are concrete and strongly-worded; however, it explicitly does so only with respect to the first two categories of international obligations—to "respect" and "protect" (par. 31 and par. 33 of the General Comment). It avoids direct references to the third category ("fulfill"), although clearly par. 34 has this category of obligations in mind. As this is the most controversial of the three categories, even more in an international realm, the Committee on Economic, Social and Cultural Rights (ESCR-Committee) was aware of the reluctance of States to acknowledge this particular international duty. They could, in the belief of many developed States, entail unforeseeable and unwanted economic and financial consequences. However, if the right to water wants be effective in its implementation, more work is needed to determine what these international obligations to fulfill exactly entail.

Apart from this lack of clarity with regard to concrete international obligations to fulfill, there are also some broader problems concerning international water obligations that derive from general international law. Three of the most crucial

[213] Alston (1991); Marong (2010); Marks (2004).

ones are the problem of necessary State consent for water assistance, the problem of "conditionality" of offered water assistance, and the problem of withdrawal of help once water assistance is initiated.

4.4.1 *Necessary Consent of Host State for International Assistance*

So far the study concentrated on international duties of foreign States under a human right to water. However, international assistance also requires something on the part of the recipient State: consent for that international water assistance. Art. 11(1) ICESCR on the right to an adequate standard of living explicitly refers to international cooperation "based on free consent". This requirement is generalizable for other human rights. It is valid as much for international assistance as it is for international cooperation; Art. 2(1) UN Charter, the equal sovereignty of States, does not allow any other conclusion.

However, at the same time, the problems with this requirement are obvious. At least in cases where there is no State authority left ("failed States"),[214] the obligation to provide humanitarian water assistance to the population of that State must remain untouched. In fact, in these cases, the obligation of other States to bridge the gap that the disappeared State authority has left open is often even more urgent, and can not undermine the obligation to fulfil: it is an obligation owed to the people, not to the State anyway. Apart from the case of "failed States", numerous governments have refused to accept international assistance, despite a massive crisis situation in their country and without the means to resolve this crisis on their own. A recent example for this kind of incident is the case of Myanmar, where in May 2008, the cyclone "Nargis" destroyed big parts of the country and its infrastructure, leaving the population, *inter alia*, without clean water. However, for several weeks, the government of Myanmar refused to accept international assistance, without accepting the true scope of the disaster and without providing sufficient assistance to their population on their own.[215] In cases like that of Myanmar, it is questionable whether the sovereignty rights of the State can outweigh the right to water of the population. At least in the immediate aftermath of a major natural disaster, States have an obligation to take the most necessary steps to provide sufficient humanitarian assistance to ensure a minimum level of water. If a State is unable to provide this assistance on its own, it must be considered part of their obligation under a right to water, to seek international assistance, and to accept it, if offered. An arbitrary refusal to accept international water assistance at the cost of the

[214] Thürer (1999); *id.* (2009); Brooks (2005).

[215] See Mydans (2008); Siemaszko (2008). See also Ford (2010); and the post-Nargis periodic reviews of the Tripartite Core Group (comprised of representatives of the government of the Union of Myanmar, the Association of Southeast Asian Nations and the UN).

affected population can, in turn, constitute a violation of the right to water of that State.

However, an arbitrary refusal must be distinguished from a reasonable refusal to accept international water assistance. States have refused in the past to accept food aid, if it was genetically modified,[216] or denied the acceptance of financial aid for the fight against human immunodeficiency virus (HIV), if some conditions tied to the financial aid threatened to undermine a public policy of that State.[217] Similar cases are feasible for the case of water: where international water assistance interferes with a public water policy of a State, or short term water assistance might damage the development of long term water perspectives, the State might be justified in refusing international aid. This will, however, be the exception and requires well-reasoned explanation by that State. The decision whether a given justification is sufficient must be made on a case-to-case basis. If a State finds itself in a temporary massive water crisis, one must certainly apply a slightly stricter standards as to which political reasons are sufficient to justify the refusal to accept vitally-important water assistance.

4.4.2 The Problem of "Conditionality"

Certainly water assistance always has "some" conditions. Firstly, all parties are obliged to respect human rights during the course of providing and accepting water assistance: preventing one human rights violation by causing another one cannot be the purpose of international water assistance. Secondly, the assistance must be used for its intended purposes and cannot be abused: the aid must reach those members of society who are without access to clean potable water.

However, international assistance, especially through IFIs is often made available only under *additional* conditions. These conditions often entail that the recipient State undertakes certain economic reforms, such as the privatization of water supply systems, as discussed in the previous chapter for the case of the city of Cochabamba.[218]

Whether such conditions are acceptable, depends largely on the purpose of the water assistance. No conditions can be justified for the granting of international water assistance if this assistance is necessary for the recipient State to meet its core obligations under the right to water. As the discussion on the philosophical foundations of water assistance has shown, for those in a position to fulfil international duties under the right to water, it is not a question of beneficence, but of obligation. If one generally owes a certain action, one has no legal or moral grounds

---

[216] See for instance Manda (2003), p. 5; and Taylor et al. (2007).

[217] See Boseley and Goldenberg (2005); and Rother (2005).

[218] See discussion at Sect. 4.2.2.3 of this book. See on the conditionality of the World Bank's aid Santiso (2001).

to impose conditions on the fulfilment of this owed action. The case becomes more difficult in circumstances where the assistance does not help to fulfil pure core obligations. If a State ensures the basic minimum of water supply to its population, but receives additional technical or financial water assistance, the given assistance is not in fulfilment of a duty, but rather beneficence. This beneficence can generally be subject to conditions. A case of such preferential treatment is for instance that of the World Trade Organization (WTO) case in *India v. EC*,[219] also known as the *GSP*-case, in which the European Community (EC) granted preferential trade tariffs to developing countries under the condition that they complied with certain drug-enforcement conditions. The WTO appellate body conceded that conditionality was generally permissible—provided it was equitable, non-arbitrary and non-discriminatory. What this case suggests for trade tariffs is also valid for the case of water assistance: helping another State to improve and develop water supply systems can be subject to conditions, if these conditions are reasonable under the factors mentioned above. Whether this is the case, has to be, again, judged on a case-by-case basis.

4.4.3 Conditions of Withdrawal and Suspension of International Water Assistance

A third question is under which conditions water assistance, after it has been started, can be withdrawn or suspended. Reasons for withdrawal or discontinuation can stem from two different spheres: that of the donor State and that of the recipient State.

Regarding the donor State, the obligation to fulfil can end when the State is no longer "in a position to do so", that is if its own situation has radically worsened (e.g. if the State itself is no longer water-abundant). More difficult is to answer is the question whether extended costs within a donor State for new other projects, such as participation in an international military intervention or new economic policies, can justify the withdrawal of international water assistance.[220] Even more doubtful is the question whether a global economic crisis, such as the most recent one, can suspend or minimize the obligations of donor States: as global crises are likely to affect the overall wealth of all nations, it may not necessarily alter significantly disparities in wealth distribution among States. In any case, should one State withdraw or suspend existing water assistance, it holds the obligation to do so in a way that does not affect the needs of the population of the recipient States more

[219] WTO Appellate Body, *European Communities—conditions for the granting of tariff preferences to developing countries*, WT/DS246/AB/R, 7 April 2004.

[220] See for instance the case of the UK withdrawing its established development assistance in Peru due to the "re-allocation" of its resources to the benefit of the reconstruction of Iraq after the invasion; see UN Commission on Human Rights, *Report submitted by the Special Rapporteur on the right of everyone to the highest attainable standard of physical and mental health, Paul Hunt: Mission to Peru*, E/CN.4/2005/51/Add.3, 4 February 2005, par. 40.

than is necessary. This might include the duty to notify the withdrawal or suspension of water assistance as early as possible and to actively get involved in the search for alternative water assistance.

However, reasons for the withdrawal of international water assistance can also lie in the sphere of the recipient State. This is the case where international assistance is no longer required in a State to realize its minimum core obligations. Withdrawing the assistance is only logical in this case if it is redirected towards other States with more severe water problems. Withdrawal must also be justified in cases in which the water assistance is evidently abused or misused by the recipient State (not an unlikely scenario given the economic value of freshwater): if it is used to commit further human rights violations (e.g. using the supplied water or improved technology to deepen an existing discrimination in terms of water access) or if the assistance is evidently not used for its purpose at all, but for personal or political gain of the government instead. Most problematic is the case in which the reason for withdrawal is the (altered) political situation in the recipient State. This includes cases in which assistance is suspended, because of continued human rights violations by the recipient State which are not directly linked to the international water assistance itself (e.g. through a new (un-democratic) government that the donor State does not want to support). This hinges upon the above mentioned problem of conditionality: can water assistance be discontinued, if (implied) conditions, such as democracy or the respect for human rights, are no longer met by a recipient State? This question certainly requires further research in the future. Its answer will depend largely on the severity of the water crisis and its relation to the demanded conditions; under no circumstances can water assistance be discontinued, if it is necessary to uphold minimum access to water for the population of the recipient State.

Altogether, international water assistance is one of the most crucial and, at the main time, contested means to fight the global water crisis. However, would a more clearly accepted human right to water in international law facilitate these international efforts?

Pessimists might argue that defining concrete international duties and deriving them from a right to water is doing a double-disservice: to the existing efforts of international water assistance on the one hand and to the right to water on the other, given the opposition that both institutions still face. However, such a line of argument can ultimately not convince. There is no alternative to tying international water assistance and the right to water together. Experience shows that hoping for States' beneficence, at least, is certainly often a "Waiting for Godot".[221]

One should instead more clearly identify what international duties the acceptance of the human right to water entails. At the same time, one must clarify what duties it does *not* entail. This would de-mystify the argumentation that the international realization of a right to water is unaffordable or endangers the wealth of the water-abundant world. International water assistance is not distributing water

[221] See the play of the same name, Beckett (2006). The main point is: Godot never comes.

wealth equally amongst nations; it aims to create basic water justice so that minimum access to water can be ensured in all countries.

Water assistance within these limits, however, must increasingly be considered a legal obligation. This chapter has shown that there is a sound philosophical cause for such obligations, be them based on consequentialist, modern Kantian or Rawlsian ideas. In particular, modern political philosophy, such as Pogge's and Risse's approaches to global justice, make a compelling case for an international duty to fulfil basic human rights standards such as a minimum level of water supply for the global population.

Will the right be implemented internationally the more clearly it is accepted? The basic necessary tools are present in the law, and a compelling moral case can be made for international water assistance. What is still lacking, as so often, is the international political will to accept this duty, even if, legally and morally, it exists already.

In this way, the human right to water and international water assistance are mutually inclusive: a human right to water can only be effectively implemented, if it also entails international means, while tying such water assistance to reliable legal claims can best be done by tying it to a right to water that is increasingly accepted in international law.

References

Ali A (2002) A conceptual framework for environmental justice based on shared but differentiated responsibilities. Research Centre for Social and Economic Research on the Global Environment (University of Norfolk). Working Paper, EDM 01-02

Alston P (1991) Revitalising United Nations work on human rights and development. Melb Univ Law Rev 18(2):216–257

Alston P (2005) Ships passing in the night: the current state of the human rights and development debate seen through the lens of the millennium development goals. Hum Rights Q 27 (3):755–829

Amor A (1998) The mandate of the UN Special Rapporteur. Emory Int Law Rev 12(2):945–950

Annan K (2006) Secretary-General's message to the third session of the Human Rights Council, Geneva, 29 November 2006. http://www.un.org/sg/statements/index.asp?nid=2333

Baderin MA, Ssenyonjo M (2010) International human rights law: six decades after the UDHR and beyond. Ashgate, Farnham

Beckett S (2006) Waiting for godot, 5th edn. Faber and Faber, London

Beisheim M (2007) Ware Wasser: Private Beteiligung bei der Wasserver- und -entsorgung in Entwicklungsländern: Lehren aus dem Fall Cochabamba. In: Rudolf B (ed) Menschenrecht Wasser? Peter Lang, Frankfurt am Main, pp 109–121

Boseley S, Goldenberg S (2005) Brazil spurns US terms for aids help. http://www.theguardian.com/world/2005/may/04/brazil.aids

Boyle A (2009) Polluter pays. In: Wolfrum R (ed) Max Planck encyclopedia of public international law. Oxford University Press, Oxford (online edition)

Branco MC, Henriques PD (2010) The political economy of the human right to water. Rev Radic Polit Econ 42(2):142–155

Brooks RE (2005) Failed states, or the state as failure. Univ Chic Law Rev 72(4):1159–1196

Buchanan A (2004) Justice, legitimacy and self-determinant, moral foundations for international law, 1st edn. Oxford University Press, Oxford
Bueno de Mesquita J, Hunt P, Khosla R (2010) The human rights responsibility of international assistance and cooperation in health. In: Gibney M, Skogly S (eds) Universal human rights and extraterritorial obligations. University of Pennsylvania Press, Philadelphia, pp 104–129
Centre for International Governance (2010) The role of the special Rapporteurs. http://www.law.leeds.ac.uk/assets/files/research/cfig/special-rapporteurs-workshop-report.pdf
Cranston M (1973) What are human rights? Bodley Head, New York
Craven M (1995) The international covenant on economic, social and cultural rights – a perspective on its development. Clarendon Press, Oxford
Dalton G (2001) Private Sector finance for water sector infrastructure: what does Cochabamba tell us about using this instrument? The School of Oriental and African Studies (SOAS), University London, Water Issues Study Group, Occasional Paper No. 37
Dupuy PM (2009) Unification rather than fragmentation of international law? The case of international investment law and human rights law. In: Dupuy PM, Francioni F, Petersmann EU (eds) Human rights in international investment law and arbitration. Oxford University Press, Oxford, pp 45–62
Fischer M, Zwetkow K (2003) Privatisierungsoptionen für den deutschen Wasserversorgungsmarkt. Zeitschrift Wasserrecht 42(3):129–156
Ford S (2010) Is the failure to respond appropriately to a natural disaster a crime against humanity? The responsibility to protect and individual criminal responsibility in the aftermath of cyclone Nargis. Denver J Int Law Policy 38(2):227–276
Gleick PH, Wolf G, Calecki EL et al (2002) The new economy of water: the risks and benefits of globalization and privatization of freshwater. Report by the Pacific Institute for Studies in Development, Environment and Security, February 2002
Glennon R (2005) Water scarcity, marketing and privatization. Tex Law Rev 83(7):1873–1902
Grimes HJ (2009) Responding to the water crisis: the complementary roles of water governance and the human right to water. J Water Law 20(2–3):119–128 (special issue: Promoting water for all)
Grotius H (2009) The Freedom of the Seas or: the right which belongs to the Dutch to take part in the East Indian Trade (transl: Ralph van Deman Magoffin), 1608 ed version. Oxford University Press, New York
Hall D, Lobina E (2007) International actors and multinational water company strategies in Europe, 1990–2003. Utilities Policy 15(2):64–77
Hall D, Lobina E, de la Motte R (2005) Public resistance to privatization in water and energy. Dev Pract 15(3–4):286–301
Hannum H (2007) Reforming the special procedures and mechanisms of the commission on human rights. Hum Rights Law Rev 7(1):73–92
Hardin G (1968) The tragedy of the commons. Science 162(3859):1243–1248
Harris PG (1999) Common but differentiated responsibility. N Y Univ Environ Law J 7(1):27–48
Harris PG (2006) Bechtel, Bolivia resolve dispute/company drops demand over water contract canceling. http://www.sfgate.com/news/article/Bechtel-Bolivia-resolve-dispute-Company-drops-2523974.php
Hey E (2011) Common but differentiated responsibilities. In: Wolfrum R (ed) Max Planck encyclopedia of public international law. Oxford University Press, Oxford (online edition)
Hill TE Jr (2006) Kantian normative ethics. In: Copp D (ed) The Oxford handbook of ethical theory. Oxford University Press, Oxford, pp 480–514
Hirsch M (2009) Investment tribunals and human rights: divergent paths. In: Dupuy PM, Francioni F, Petersmann EU (eds) Human rights in international investment law and arbitration. Oxford University Press, Oxford, pp 97–115
Human Rights Council (HRC) (2008) Manual of operations of the special procedures of the Human Rights Council. http://www.ohchr.org/Documents/HRBodies/SP/Manual_Operations2008.pdf
Irmscher R (2001) Kommunale Wasserwirtschaft in einem sich ändernden Umfeld. In: Büscher E (ed) Wasserwirtschaft im Aufbruch. Luchterhand, Cologne

Jeffords C, Shah F (2012) On the natural and economic difficulties to fulfilling the human right to water within a neoclassical economics framework. http://www.icape.org/h3-jeffords-shah.pdf

Knoll-Tudor I (2009) The fair and equitable treatment standard and human rights norms. In: Dupuy PM, Francioni F, Petersmann EU (eds) Human rights in international investment law and arbitration. Oxford University Press, Oxford, pp 310–343

Laskowski SR (2010) Das Menschenrecht auf Wasser: Die rechtlichen Vorgaben zur Sicherung der Grundversorgung mit Wasser und Sanitärleistungen im Rahmen einer ökologisch-nachhaltigen. Mohr Siebeck, Tübingen

Leckie S, Gallagher A (2006) Economic, social, and cultural rights: a legal resource guide. University of Pennsylvania Press, Philadelphia

Lever P (2007) Wasser: Die Global Players und ihre Strategien – Privatisierung als Fluch oder Segen? In: Rudolf B (ed) Menschenrecht Wasser. Peter Lang, Frankfurt am Main, pp 85–92

Lobina E (2000) Cochabamba: water war. Public Services International Research Unit (PSIRU). Working Paper Series at Business School of University of Greenwich

Locke J (1821) Two treatises on government: book II. Whitmore and Fenn, London

Mamlyuk BN (2009) Analyzing the polluter pays principle through law and economics. Southeast Environ Law J 18(1):39–80

Manda LO (2003) Controversy rages over "GM" food aid. Africa Renewal 16(4):5

Marks S (2004) The human right to development: between rhetoric and reality. Harv Hum Rights J 17:137–168

Marong A (2010) Development, right to, international protection. In: Wolfrum R (ed) Max Planck encyclopedia of public international law. Oxford University Press, Oxford (online edition)

Mayeda G (2008) International investment agreements between developed and developing countries: dancing with the devil? A case commentary on the Vivendi, Sempra and Enron awards. McGill Int J Sustain Dev Law Policy 4(2):189–230

McFarland Sánchez-Moreno M, Higgins T (2004) No recourse: transnational corporations and the protection of economic, social and cultural rights in Bolivia. Fordham Int Law J 27 (5):1663–1805

Mydans S (2008) Myanmar faces pressure to allow major aid effort. http://www.nytimes.com/2008/05/08/world/asia/08myanmar.html?_r=1&hp

Naegele J (2004) What is wrong with full-fledged water privatization? J Law Soc Chall 6:99–130

Norris AL, Metzidakis KE (2010) Public protests, private contracts: confidentiality in ICSID arbitration and the Cochabamba water war. Harv Negot Law Rev 15(1):31–76

O'Neill O (1980) Kantian approaches to some famine problems. In: Regan T (ed) Matters of life and death. Random House, New York, pp 285–294

Office of the High Commissioner for Human Rights (OHCHR) (2001) Fact Sheet No. 27: Seventeen frequently asked questions about United Nations Special Rapporteurs. http://www.ohchr.org/Documents/Publications/FactSheet27en.pdf

Office of the High Commissioner for Human Rights (OHCHR) (2005) Economic, social and cultural rights: handbook for national human rights institutions. United Nations, New York/Geneva

Pahl-Wostl C, Gupta J, Petry D (2008) Governance and the global water system: a theoretical exploration. Glob Gov 14(4):419–436

Perin M (2000) Azurix water bugs Argentina. http://houston.bizjournals.com/houston/stories/2000/05/08/tidbits.html

Petersmann EU (2009) Constitutional theories of international economic adjudication and investor-state arbitration. In: Dupuy PM, Francioni F, Petersmann EU (eds) Human rights in international investment law and arbitration. Oxford University Press, Oxford, pp 138–194

Pindyck RS, Rubinfeld DL (2009) Microeconomics, 7th edn. Pearson Education, Upper Saddle River

Pinheiro PS (2003) Global insights: musings of a UN Special Rapporteur. Glob Gov 9(7):7–13

Pogge TW (2002) World poverty and human rights. Polity Press/Blackwell Publishing Ltd, Oxford

Rawls J (1958) Justice as fairness. Philos Rev 67(2):164–194

Rawls J (1971) A theory of justice, 1st edn. Harvard University Press, Cambridge
Reed L et al (2011) Guide to ICSID arbitration, 2nd edn. Kluwer Law International, Alphen aan den Rijn
Riedel E (2006) The human right to water and general comment no. 15 of the CESCR. In: Riedel E, Rothen P (eds) The human right to water. Berliner Wissenschafts-Verlag, Berlin, pp 19–36
Riedel E (2011) International covenant on economic, social and cultural rights (1966). In: Wolfrum R (ed) Max Planck encyclopedia of public international law. Oxford University Press, Oxford (online edition)
Risse M (2008a) Original Ownership of the Earth: a contemporary approach. Harvard Kennedy School Working Paper Series, Cambridge
Risse M (2008b) Immigration and common ownership of the Earth. In: Conference paper for United Nations Development Programme, Cambridge, December 2008
Risse M (2012) On global justice. Princeton University Press, Princeton
Risse M (2013) The human right to water and common ownership of the Earth, HKS Faculty Research Working Paper Series, RWP13-003, February 2013
Rother L (2005) Prostitution puts U.S. and Brazil at odds on AIDS policy. http://www.nytimes.com/2005/07/24/international/americas/24brazil.html
Santiso C (2001) Good governance and aid effectiveness: the World Bank and conditionality. Georget Public Policy Rev 7(1):1–22
Scheinin M (2009) Towards a world court of human rights. Research report to the Swiss initiative to commemorate the 60th anniversary of the Universal Declaration of Human Rights. http://www.udhr60.ch/report/hrCourt_scheinin0609.pdf
Schill S (2006) Fair and equitable treatment under investment treaties as an embodiment of the rule of law. Transnatl Dispute Manag 3(5):9–30
Selden J (1636) Mare Clausum Seu De Dominio Maris. Maire, Lugduni Batavorum
Sen A (1987) On ethics and economy. Blackwell, Oxford
Sepúlveda M (2006) Obligations of international assistance and cooperation in an optional protocol to the international covenant on economic, social and cultural rights. Neth Q Hum Rights 24(2):7–56
Shaw W (2006) The consequentialist perspective. In: Dreier J (ed) Contemporary debates in moral theory. Wiley-Blackwell, Malden, pp 5–20
Shue H (1980) Basic rights: subsistence, affluence, and U.S. foreign policy. Princeton University Press, Princeton
Siemaszko C (2008) Myanmar government refuses U.S. aid. http://www.nydailynews.com/news/world/myanmar-government-refuses-u-s-aid-article-1.328655
Singer P (1972) Famine, affluence, and morality. Philos Public Aff 1(1):229–243
Singer P (2004) One world: the ethics of globalization, 2nd edn. Yale Universty Press, New Haven
Skogly SI (2006) Beyond national borders: states' human rights obligations in international cooperation. Intersentia, Antwerp
Smith A (1790) The theory of moral sentiments, 6th edn. A. Miller, London
Solanes M, Jouravlev A (2006) Water governance for development and sustainability. CEPAL – Serie Recursos naturals e infraestructura No. 111. United Nations, Santiago
Steiner HJ, Alston P, Goodman R (2008) International human rights in context, law, politics, morals – text and materials, 3rd edn. Oxford University Press, New York
Taylor PL, Asworth DN, Kitchen DE (2007) Helping Africa Achieve Millennium Development Goal 1: a hyperbole-free US genetically modified food aid policy. Africa Policy Journal, Harvard Kennedy School of Government, 03/2007
Thielbörger P (2009) The human right to water versus investor rights: double-dilemma or pseudo-conflict? In: Dupuy PM, Francioni F, Petersmann EU (eds) Human rights in international investment law and arbitration. Oxford University Press, Oxford, pp 487–511

Thielbörger P (2013) Governing international watercourses: implications of the human right to water, chapter 3. In: Kibaroglu A, Kirschner AJ, Mehring S, Wolfrum R (eds) Water Law and Cooperation in the Euphrates Tigris region: a comparative and inter-disciplinary study of international and national water law. Brill, Leiden

Thürer D (1999) The "failed state" and international law. Int Rev Red Cross 81(836):731–761

Thürer D (2009) Failing states. In: Wolfrum R (ed) Max Planck encyclopedia of public international law. Oxford University Press, Oxford (online edition)

Trechsel S (2003) A World Court for Human Rights? Northwest Univ J Int Hum Rights 1:i–xviii, art 3

Tullis P (2010) The U.N.'s death squad watchdog. http://www.psmag.com/legal-affairs/the-u-n-s-death-squad-watchdog-11440

Wälde TW (2004) ICSID 'Annulment Committee'. Transnatl Dispute Settlement 1(1). www.transnational-dispute-management.com/article.asp?key=57

Welwood W (1613) An abridgement of all sea-lawes. Man Publishing, London

Welwood W (1615) De dominio maris. Eld Publishing, London

Werner J (2009) Limits of commercial investor-state arbitration: the need for appellate review. In: Dupuy PM, Francioni F, Petersmann EU (eds) Human rights in international investment law and arbitration. Oxford University Press, Oxford, pp 115–117

Woodhouse EJ (2003) The Guerra del Agua and the Cochabamba concession: social risk and foreign direct investment in public infrastructure. Stanford J Int Law 39(2):295–350

Young IM (2004) Responsibility and global labor justice. J Polit Philos 12(4):365–388

Chapter 5
Conclusion

In the course of the last years, the human right to water has been largely discussed in international law, international politics and civil society. Since 1992, every 22 March is celebrated as annual "world water day".[1] The United Nations (UN) named the entire decade 2005–2015 as "The International Decade for Action: Water for Life".[2] Famously, the UN Millennium Development Goals demand that the proportion of people living in hunger and thirst be halved by 2015. Since 1997, six world water fora have been held, reaffirming on a tri-annual basis that water is one of the most important common goods, and that no person should live deprived of it.[3] Since 2008, Catarina de Albuquerque holds the mandate of the Independent Expert, now called Special Rapporteur, to monitor whether States obey their human rights obligations related to access to safe drinking water and sanitation.[4] It is appropriate to state that water has become one of the favorite human rights topics of many western governments and numerous NGOs, and most certainly of the United Nations. Dazzled by all of this "water-mania", one is easily tempted to forget an important fact: that currently there is not easily a self-standing, comprehensive and at the same time legally binding human right to water in international law. Nevertheless, it can still be said that a human right to water exists in international law as well as in most national legal orders—it is just very distinctive and very fragmented. One could say it is a right of its very own kind.

The second chapter of this book has identified the different ways of protecting a right to water in different legal systems, be it in the form of a constitutional right, in the form of ordinary rights or through more broad concepts like that of "services

*All cited webpages were last accessed on 1 August 2013.

[1] For more information on the annual celebrations, see http://www.worldwaterday.org/.

[2] See UNGA, *International Decade for Action, "Water for Life", 2005–2015*, A/RES/58/217, 23 December 2003.

[3] Webpage of the 6th World Water Forum, Marseille 2012, including history of the five previous ones, http://www.worldwaterforum6.org.

[4] See chapter on the monitoring of the mandate at Sect. 4.1 of this book.

P. Thielbörger, *The Right(s) to Water*,
DOI 10.1007/978-3-642-33908-0_5, © Springer-Verlag Berlin Heidelberg 2014

of general interest". In order to protect basic water needs, most legislators and judiciaries have developed approaches that are consistent with the systems they operate in, influenced by the water needs the communities in these systems face. The tradition of many developed States goes in the direction to view water from the lens of a purely economic and social right, as water disputes in these countries tend to arise on questions of water quality and water prices. Less developed countries, such as India, on the other hand, have sometimes chosen approaches that focus on those elements of the right that are essential for bare survival. It is remarkable what all systems have in common: judges were one of the driving forces for the development of the right. Legislators have been more hesitant about establishing rights that (supposedly) entail significant financial commitments.

The fact that these national and regional approaches are so diverse is neither a problem nor is it a surprise. It is rather a reflection of the different regional water situations and of different traditions of how to balance individual needs and State obligations. Some legal regimes are very restrictive in their acceptance of individual basic rights; others have accepted a variety of individual rights, but have in the meantime decided to put a hold on the "inflation" of rights; again others hold the tradition of putting a strong emphasis on civil and political rights, while remaining skeptical about the acceptance of economic and social rights.

Amongst other factors, these very different situations in terms of water scarcity, paired with the different traditions in guaranteeing individual entitlements, have made it difficult in the past to agree upon one comprehensive legal approach to water in international law. It is rather the expression of the agreement to disagree, and to differ. This is one of the reasons why the enumeration of the right to water in international agreements is more erratic than systematic. General Comment No. 15, the most important document in international law on the topic of individual water needs, has not changed this situation. It is the smallest common denominator of what a human right to water should mean. It describes in a logically coherent and politically wise way possibilities as to how the normative content of a human right to water should be understood. It makes a compelling case of how one cannot accept certain other human rights without guaranteeing access to safe drinking water. It should therefore be more understood as a logical exercise than as an assessment of an existing legal situation. It is a tool to facilitate the interpretation of other existing human rights treaties.

Thus, General Comment No. 15 is not the *legal* break-through that it has sometimes been claimed to be. The previous dilemma—that there is no clearly defined and at the same time comprehensive and legally binding recognition of a human right to water under international law—was not solved by the General Comment: it lacks legal-bindingness. However, the value of General Comment No. 15 could not be over-estimated. Never before, and never after, has a document in larger detail elaborated on the content, the relating obligations and the possible legal foundations of a human right to water. In terms of its political significance, it has been, and still is, of utmost importance. General Comment No. 15 has been criticized too harshly in the past by some scholars—it has done what it could, while resisting the temptation to try and do more than it should. It has not proscribed a duty of the States to accept a human right to water. It has rather made a logical case of what a human right to water could look like and tried to collect evidence that, in fact, most of the underlying elements of such a right, including individual entitlements, are already protected through other existing human rights. In terms of its *political* meaning, General Comment No. 15 must be considered as one of the most influential and successful of all General Comments.

In its third chapter, this book has posed the question whether an attempt to approach basic water needs through the legal means of a human right is a logically coherent endeavor. By employing Descartes' "method of doubt", it has examined conceptual possible objections towards such an approach and asked whether and how these obstacles can be overcome. These are basic questions that have too often been neglected. The louder the political claim for a human right to water grew, the less were systematic objections to the concept of a right to water discussed. However, he who wants to make the stable case *for* a human right to water must show that a convincing case against it cannot be made. This chapter has in particular addressed three sub-questions: Is *water* a suitable content of a right at all? Is a *right* the best approach to address the human basic need for water? And would such a right then qualify as a *human* right?

The study has shown that access to water is a suitable object for the content of a human right. I have argued that in the middle of the twentieth century, at the time the most important human rights instruments were negotiated, water was not yet conceived as a scarce resource in the way it is today. The approach of phrasing access to limited natural goods in terms of individual entitlements had not ripened by then. With this mindset rapidly shifting and water scarcity at the same time reaching a new peak, however, the time to seriously discuss water in human right terms has now come. Letting time go by, would mean to miss the biggest chance to establish such a right.

Secondly, the chapter has shown that a right to water is not the only, but a plausible legal means to approach the basic human need for water. Creating such a right would not mean the establishment of an "impossible" right—a category of rights one should indeed strive to avoid. It would rather mean accepting a right that is currently still difficult and costly to realize. However, as we know from the philosophy of rights—notably for instance from MacCormick's famous example on children's rights—rights are especially needed where their realization is not effortless. To allow the argument that a need which is hard to realize should better not be phrased in terms of a right, would lead the general meaning of "rights" to become absurd. Phrasing access to water in terms of a right has also the advantage of putting the individual into a moral position that is not inferior than that one of the State, thereby creating a feasible added value for the rights-based approach over any other approaches.

Thirdly, such a right to water would also clearly fall into that very special category of *human* rights. The dispute on what human rights are is almost as old as the term itself. However, regardless of what human rights school one follows (the most prominent being probably the natural school and deliberative school), it is very hard to argue that water, if accepted as a right, would not qualify as a human right. What could be *human*, if not water?

However, stating that a human right to water can be accepted in a logically coherent manner, does not yet answer the question of what this recognition should look like. I have argued that, like hardly any other right, the right to water is a right that combines very different legal entitlements. Due to the enormous impact of General Comment No. 15, some of these aspects have recently sunk into oblivion. It is a right that in its nucleus protects human life and dignity—no human being can survive for longer than a few days without drinking water. It is, however, at the same time derived from several socio-economic rights: to realize the highest attainable level of health and allow for an adequate standard of living, is simply not possible without clean and affordable drinking water. Although to a certain degree this "hybrid nature" can be found in many human rights—they have a (more essential) core and a (more extensive) mantle—it is particularly striking in the case of the human right to water. This is one of the reasons why for a long time there was no agreement between those who wanted to assign the right the status of a civil and political right (by now considered an "old-fashioned" position), and those who saw it in line with other economic and social rights (by now the largely prevailing position). This study has suggested to positively acknowledge the hybrid nature of the right to water, as such acknowledgement has positive effects for the right's recognition. Those States that have not ratified the ICESCR, but the ICCPR, should be reminded that they are also bound by the nucleus of the right to water. This nucleus must be subject to the individual communication procedure of the ICCPR; a procedure whose equivalent is, given the new Optional Protocol of the ICESCR, on its way, but for the foreseeable future less effective than its ICCPR counterpart. Thus, this study has not suggested to move back and attach the right to water to the right to life. It has rather suggested to embrace the right in its fragmented nature as both civil-political *and* socio-economic. What has rightly been suggested as "the core" for all established economic and social rights, can for the right to water as easily be derived from the right to life, thereby making the right even stronger in some ways. In this sense, the not clearly defined status of the right to water is also an opportunity: recognizing it as a right with different layers allows us to make a virtue out of this necessity. These considerations have also inspired the title of this book, speaking rather of the right(s) to water than of only one single right.

The study has also suggested a clearer distinction between the right to water and the right to sanitation. Albeit both rights are increasingly mentioned together, they are distinctly different in many ways. Although recent figures provided by the UN suggest that today insufficient access to sanitation is in the meantime an even bigger problem than lacking access to drinking water, and therefore sanitation deserves international attention, the right to water and the right to sanitation should not be treated as one. Firstly, amongst States, support for the right to water is still more wide-spread than support for the right to sanitation. More importantly, the two rights, although closely related, have different parent rights from which they are derived: whereas the right to water in its nucleus has its strongest link to the right to life, the right to sanitation is more closely related to the idea of human dignity. To assume that two rights could be summarized as one, would contradict the mechanism by which both of them were created: the idea of legal derivation. Legal derivation means that the derivative right carries the characteristics of the parent right, and, thus, rights with different parent rights cannot be identical. However, stating that both rights are not identical does not mean claiming that they cannot be treated jointly in international declarations and other political instruments, if the context and purpose of the documents do allow so. There is in fact often good reason to treat them jointly; this is just not the same as claiming they are the same.

Finally, this book has also scrutinized the relationship between the different rights to water: those ones identified in the first chapter in the different national and regional orders, and the human right to water as part of international law. It uses the ECJ's famous *Kadi* judgment, in which the court ultimately examined the relation and hierarchy between European law and international law, to draw conclusions also for the case of national and international law. The study identifies two different approaches to the problem: either to accept the general primacy of international law, in particular the UN Charter (as suggested by the Court of First Instance in the *Kadi*-case), or to allow primacy of international law over national secondary law, while still ensuring the protection of rights through upholding national primary law (as suggested by the European Court of Justice (ECJ) for the case of European law). This study has generally supported the approach taken by the ECJ, but suggested a slightly attenuated approach to the one adopted in *Kadi*. If the level of protection of the right to water in a national legal system undergoes the standards demanded by international human rights law—consider for instance the permission for water access to be cut after continued non-payment—the ECJ's approach yields a reasonable result: national primary law must be the ultimate benchmark for States' internal actions. States are then called upon to bring their constitutions as soon as possible in line with their international obligations. If the norms conflicting with the human right to water, however, are rooted in domestic ordinary law, the human right requirements must prevail directly. However, if the level of national protection exceeds the level of international protection—consider for instance a ban on water privatization or a large amount of basic water supply for each household for free—it can make no difference whether this extensive protection of the right is guaranteed in domestic primary or secondary law. The human right to water should therefore be understood as a guide for States as to what minimum standards must be met. To what extent States go beyond this minimum standard, and through which means the realization of such a more extensive guarantee of the right can best be met in line with national legal and cultural traditions, remains entirely in the decision of the States themselves.

However, saying that one can establish a human right to water, and identifying the way that would be most consistent with existing national and regional approaches, is different from saying that one ought to do so. It would be too easy to simply assume that the establishment of a human right to water on its own would contribute significantly to the solution of the global fresh water crisis. On its own, a human right to water would most likely be doomed to be a toothless tiger. Research indicates that structures of good water governance in fact make a big difference for the effective fight against global thirst, and that, in turn, a discourse on the right to water is beneficial for good water governance structures. Emphasizing that it is not

States' benevolence, but the individual's entitlement to have access to safe water, is important for the development of structures of good water governance. The fourth chapter of this book has addressed exactly this disillusioning thought: that the recognition of the right to water on its own might not be the silver bullet to solve the global water crisis. The chapter therefore identifies some elements that would make the right more effective in practice. As for every right, in particular rights with a strong economic and social dimension, it is independent monitoring, enforcement also against the private sector and international realization that can account for such a difference. The discussion on the right to water should therefore concentrate on finding ways of how to better monitor, realize and enforce compliance with human rights obligations related to access to safe drinking water.

In terms of independent monitoring, great progress has recently been achieved. The introduction of the mandate of the Independent Expert on the issue of human rights obligations related to access to safe drinking water and sanitation, currently held by Catarina de Albuquerque and renamed as Special Rapporteur on the human right to water and sanitation in 2011, has largely improved the effective monitoring of human rights violations related to water and sanitation. The mandate has highlighted successfully forgotten topics as well as the most contested issues with respect to a human right to water. It has also undertaken a variety of country studies so that one can talk of a truly global monitoring procedure. However, there are serious problems that remain to be solved, such as dramatic understaffing (a typical problem of UN Special Procedures), the requirement of State consent for getting permission to conduct country visits, and the lack of follow-up monitoring mechanisms once recommendations have been made by the Expert.

In terms of enforcing the right towards the private sector, Chap. 4 has pointed out the general tensions between the market principle as basis of privatization, and the idea of a human right to water, in particular the fact that water suppliers usually hold natural monopolies in a given market, that the "desired" market equilibrium is incompatible with the ideal of universal access, and that water cannot easily be treated as a private good, if accepted as a human right. Three case studies of arbitral cases before the International Centre for the Settlement of Investment Disputes (ICSID), the most common instance to decide on disputes between investors and States after failed water privatizations, have shown that it is often the investors and the international corporations that are successful, because the right to water of individuals is often not part of the applicable law before these courts. This study has suggested a re-thinking of the "applicable law" before the ICSID, in order to ensure that basic human rights standards relating to water access are upheld in all arbitral decisions. If water is clearly accepted as one of these human rights, arbitrators would only need to decide whether to uphold this minimum standard—not whether the right to water was part of it.

In terms of realizing the right to water, attention is very often given to the national level. This is understandable given that territory and jurisdiction are generally restricting the applicability of human rights. However, the national level is not the only relevant level to take into focus. The global water crisis will not be solved without intensive international water assistance. The study suggests that there are sound philosophical reasons to assume that solving the water problems of the poorest countries is also a responsibility of the international community. Apart from traditional approaches in moral philosophy, these include notably some modern approaches to global justice and common ownership of the earth, as articulated for instance by Thomas Pogge and Mathias Risse. There is also a variety of possible legal means at hand of how to overcome the general assumption of human rights being tied to territory and jurisdiction alone. These possibilities are currently simply not used sufficiently due to a lack of political will amongst the developed States. Behind this lack of political will lies the—largely false—assumption that international water assistance aims to equalize water standards globally. In fact, international water obligations only aim to achieve the guarantee of the very basic fulfillment of the human right to water on a global scale.

Will the immediate future hold a chance for these changes? At least, progress is visible. It is most encouraging that the UN General Assembly (UNGA) resolution of July 2010 and the HRC resolution of September 2010 were both adopted without explicitly dissenting votes; the latter even without a vote. This is also clear indication for a norm of international custom that is on the rise. These developments give reason for the hope that in the foreseeable future water will be accepted as human right in a legally binding, comprehensive and self-standing manner. Until then, understanding the right to water in the suggested way, as a right of a hybrid nature with different components, maximizes the effect that the right can have already today, and is the best description of the legal claims the right in fact combines.

Altogether, the discussion must move away from the bare discussion of the existence or non-existence of a right and shift its focus onto three questions: Firstly, how do we effectively identify violations of a human right to water? Secondly, how do we develop policies to improve compliance with a human right to water? And thirdly, what mechanisms can be put in place to counter those that still do not comply with the obligations under such a human right to water? This book has started to give some answers to these questions and hopes for further problem-solving oriented discussion on these issues. If these questions are successfully addressed, but only then, can a human right to water be the so urgently needed answer to one of the most pressing questions—that one of the global water crisis—that humankind will have to face in the twenty-first century.

List of Documents

1. United Nations Documents

1.1 United Nations General Assembly (UNGA)

UNGA, *Universal Declaration of Human Rights*, Res. 217 A (III), A/810 at 71, 10 December 1948

UNGA, *Declaration on Principles of International Law concerning Friendly Relations and Co-operation among States in accordance with the Charter to the United Nations*, Res. 2625 (XXV), A/8028, 24 October 1970

UNGA, *The right to development*, A/RES/54/175, 15 February 2000

UNGA, *United Nations Millennium Declaration*, A/RES/55/2, 18 September 2000

UNGA, *Articles on the Responsibility of States for Internationally Wrongful Acts*, A/RES/56/83, 28 January 2002

UNGA, *International Decade for Action, "Water for Life", 2005–2015*, A/RES/58/217, 23 December 2003

UNGA, *Human Rights Council*, A/RES/60/251, 3 April 2006

UNGA, *Convention on the Rights of Persons with Disabilities*, A/RES/61/106, 13 December 2006

UNGA, *International Year of Sanitation, 2008*, A/RES/61/192, 6 February 2007

UNGA, *Third Committee recommends General Assembly adoption of optional protocol to International Convention on Economic, Social and Cultural Rights*, Press Release GA/SHC/3938, 18 November 2008

UNGA, *Optional Protocol to the International Covenant on Economic, Social and Cultural Rights*, A/RES/63/117, 5 March 2009

UNGA, *Summary of the 108th plenary meeting of the 64th General Assembly*, A/64/PV.108, 28 July 2010

UNGA, *General Assembly adopts resolution recognizing access to clean water, sanitation as human right*, Press Release GA/10967, 28 July 2010

DOI 10.1007/978-3-642-33908-0,

UNGA, *The human right to water and sanitation*, A/RES/64/292, 3 August 2010

UNGA, *Human rights obligations related to access to safe drinking water and sanitation*, A/65/254, 6 August 2010

UNGA, *Keeping the promise: united to achieve the Millennium Development Goals*, A/RES/65/1, 19 October 2010

UNGA, *Integrating non-discrimination and equality into the post-2015 development agenda for water, sanitation and hygiene*, A/67/270, 8 August 2012

1.2 United Nations Commission on Human Rights

UN Commission on Human Rights, *The Limburg Principles on the Implementation of the International Covenant on Economic, Social and Cultural Rights*, E/CN.4/1987/17, 8 January 1987, Annex

UN Commission on Human Rights, *Question of the realization in all countries of the economic, social and cultural rights contained in the Universal Declaration of Human Rights and in the International Covenant on Economic, Social and Cultural Rights, and study of special problems which the developing countries face in their efforts to achieve these human rights*, E/CN.4/RES/2003/18, 22 April 2003

UN Commission on Human Rights, *Report of the Special Rapporteur on adequate housing as a component of the right to an adequate standard of living, Miloon Kothari*, E/CN.4/2004/48, 8 March 2004

UN Commission on Human Rights, *Report submitted by the Special Rapporteur on the right of everyone to the highest attainable standard of physical and mental health, Paul Hunt: Mission to Peru*, E/CN.4/2005/51/Add.3, 4 February 2005

1.3 Human Rights Committee (CCPR-Committee)

CCPR-Committee, *E.H.P. v. Canada*, Communication No. 67/1980, CCPR/C/17/D/67/1980, 27 October 1982

CCPR-Committee, *General Comment No. 3: Implementation at the national level (Art. 2)*, HRI/GEN/1/Rev.9 (Vol. I), 27 May 2008, at 174

CCPR-Committee, *General Comment No. 6: The right to life (Art. 6)*, HRI/GEN/1/Rev.9 (Vol. I), 27 May 2008, at 176

1.4 United Nations Human Rights Council (HRC)

HRC, *Report of the Working Group on the Right to Development on its eighth session*, A/HRC/4/47, 14 March 2007

HRC, *Institution building of the United Nations Human Rights Council*, A/HRC/RES/5/1, 7 August 2007, Annex

HRC, *Human rights and access to safe drinking water and sanitation*, A/HRC/RES/7/22, adopted on 28 March 2008

HRC, *Report of the independent expert on the issue of human rights obligations related to access to safe drinking water and sanitation, Catarina de Albuquerque: Mission to Costa Rica*, A/HRC/12/24/Add.1, 23 June 2009

HRC, *Report of the independent expert on the issue of human rights obligations related to access to safe drinking water and sanitation, Catarina de Albuquerque*, A/HRC/12/24, 1 July 2009

HRC, *Report of the independent expert on the issue of human rights obligations related to access to safe drinking water and sanitation*, A/HRC/15/31, 29 June 2010

HRC, *Report of the independent expert on the issue of human rights obligations related to access to safe drinking water and sanitation, Catarina de Albuquerque: Mission to Egypt*, A/HRC/15/31/Add.3, 5 July 2010

HRC, *Joint report of the independent expert on the question of human rights and extreme poverty, Magdalena Sepúlveda Cardona, and the independent expert on the issue of human rights obligations related to access to safe drinking water and sanitation, Catarina de Albuquerque: Mission to Bangladesh*, A/HRC/15/55, 22 July 2010

HRC, *Human rights and access to safe drinking water and sanitation*, A/HRC/RES/15/9, 6 October 2010

HRC, *The human right to safe drinking water and sanitation*, A/HRC/RES/16/2, 8 April 2011

HRC, *Report of the Special Rapporteur on the human right to safe drinking water and sanitation, Catarina de Albuquerque: Compilation of good practices*, A/HRC/18/33/Add.1, 29 June 2011

HRC, *Report of the Special Rapporteur on the human right to safe drinking water and sanitation, Catarina de Albuquerque: Mission to Slovenia*, A/HRC/18/33/Add.2, 4 July 2011

HRC, *Report of the Special Rapporteur on the human right to safe drinking water and sanitation, Catarina de Albuquerque: Mission to Japan*, A/HRC/18/33/Add.3, 4 July 2011

HRC, *Report of the Special Rapporteur on the human right to safe drinking water and sanitation, Catarina de Albuquerque: Mission to the United States of America*, A/HRC/18/33/Add.4, 2 August 2011

HRC, *Report of the Special Rapporteur on the human right to safe drinking water and sanitation, Catarina de Albuquerque: Mission to Namibia*, A/HRC/21/42/Add.3, 28 June 2012

HRC, *Stigma and the realization of the human rights to water and sanitation*, A/HRC/21/42, 2 July 2012

HRC, *Report of the Special Rapporteur on the human right to safe drinking water and sanitation, Catarina de Albuquerque: Mission to Uruguay*, A/HRC/21/42/Add.2, 2 July 2012

HRC, *Report of the Special Rapporteur on the human right to safe drinking water and sanitation, Catarina De Albuquerque: Mission to Senegal*, A/HRC/21/42/Add.1, 16 August 2012

1.5 Committee on Economic, Social and Cultural Rights (ESCR-Committee)

ESCR-Committee, *Report of the open-ended working group to consider options regarding the elaboration of an optional protocol to the International Covenant on Economic, Social and Cultural Rights on its second session*, E/CN.4/2005/52, 10 February 2005

ESCR-Committee, *General Comment No. 3: The nature of States parties' obligations (Art. 2 par. 1)*, HRI/GEN/1/Rev.9 (Vol. I), 27 May 2008, at 7

ESCR-Committee, *General Comment No. 4: The right to adequate housing (Art. 11 par. 1)*, HRI/GEN/1/Rev.9 (Vol. I), 27 May 2008, at 11

ESCR-Committee, *General Comment No. 12: The right to adequate food (Art. 11)*, HRI/GEN/1/Rev.9 (Vol. I), 27 May 2008, at 55

ESCR-Committee, *General Comment No. 14: The right to the highest attainable standard of health (Art. 12)*, HRI/GEN/1/Rev.9 (Vol. I), 27 May 2008, at 78

ESCR-Committee, *General Comment No. 15: The right to water (Art. 11 and 12)*, HRI/GEN/1/Rev.9 (Vol. I), 27 May 2008, at 97

ESCR-Committee, *General Comment No. 19: The right to social security (Art. 9)*, HRI/GEN/1/Rev.9 (Vol. I), 27 May 2008, at 152

ESCR-Committee, *General Comment No. 20: Non-discrimination in economic, social and cultural rights (Art. 2 par. 2)*, E/C.12/GC/20, 2 July 2009

ESCR-Committee, *General Comment No. 21: Right of everyone to take part in cultural life (Art. 15 par. 1(a))*, E/C.12/GC/21, 21 December 2009

1.6 United Nations Conference Reports

United Nations Conference on the Human Environment, *Declaration on the United Nations Conference on the Human Environment*, A/CONF.48/14/Rev.1, at 3, 1973, adopted on 16 June 1972

Report of the UN Water Conference, Mar del Plata, 14–25 March 1977, E/CONF.70/29, 1977

UN Conference on Environment and Development, Rio de Janeiro, 3–14 June 1992, *Agenda 21*, http://www.unep.org/Documents.Multilingual/Default.asp?documentid=52

UNGA, *Report of the United Nations Conference on Environment and Development: Rio Declaration on Environment and Development*, A/CONF.151/26 (Vol. I), 12 August 1992, Annex I

World Conference on Human Rights, *Vienna Declaration and Programme of Action*, A/CONF.157/23, 12 July 1993

Report of the International Conference on Population and Development, Cairo, 5–13 September 1994, A/CONF.171/13, 18 October 1994

1.7 Other Bodies

ECOSOC, *International Covenant on Economic, Social and Cultural Rights*, E/RES/1987/5, adopted on 26 May 1987

ECOSOC, *International Covenants on Human Rights*, E/RES/1990/45, adopted on 25 May 1990

Committee on the Rights of Child, *Day of General Discussion on "Resources for the Rights of the Child—Responsibility of States"*, 5 October 2007, http://www2.ohchr.org/english/bodies/crc/docs/discussion/recommendations2007.doc

2. Jurisprudence/Case Law

2.1 International Courts

2.1.1 International Court of Justice (ICJ)

ICJ, *Right of Passage over Indian Territory* (Portugal v. India), Judgment of 12 April 1960, ICJ Reports (1960), 6

ICJ, *Temple of Preah Vihear* (Cambodia v. Thailand), Judgment of 15 June 1962, ICJ Reports (1962), 6

ICJ, *North Sea Continental Shelf* (Federal Republic of Germany v. Denmark; Federal Republic of Germany v. Netherlands), Judgment of 20 February 1969, ICJ Reports (1969), 3

ICJ, *Delimitation of the Maritime Boundary in the Gulf of Maine Area* (Canada v. USA), Judgment of 12 October 1984, ICJ Reports (1984), 246

ICJ, *Case Concerning the Military and Paramilitary Activities In and Against Nicaragua* (Nicaragua v. USA), Judgment of 27 June 1986, ICJ Reports (1986), 14

ICJ, *Legality of the Threat or Use of Nuclear Weapons*, Advisory Opinion of 8 July 1996, ICJ Reports (1996), 226

ICJ, *Gabčíkovo-Nagymaros Project* (Hungary v. Slovakia), Judgment of 25 September 1997, ICJ Reports (1997), 7

ICJ, *Gabčíkovo-Nagymaros Project* (Hungary v. Slovakia), Judgment of 25 September 1997, Separate Opinion of Vice-President Christopher Gregory Weeramantry, ICJ Reports (1997), 88
ICJ, *Case concerning Pulp Mills on the River Uruguay* (Argentina v. Uruguay), Order on the Request for the Indication of Provisional Measures of 13 July 2006, ICJ Reports (2006), 113
ICJ, *Case concerning Pulp Mills on the River Uruguay* (Argentina v. Uruguay), Judgment of 20 April 2010, ICJ Reports (2010), 14

2.1.2 European Court of Justice (ECJ) and Court of First Instance (CFI)

CFI, Case T-315/01, *Yassin Abdullah Kadi v. Council of the European Union and Commission of the European Communities*, Judgment of the Court of First Instance of 21 September 2005, [2005] ECR II-3649
ECJ, Joined Cases C-402/05 P and C-415/05 P, *Yassin Abdullah Kadi and Al Barakaat International Foundation v. Council of the European Union and Commission of the European Communities*, Opinion of Advocate General Maduro of 23 January 2008, [2008] ECR I-6351
ECJ, Joined Cases C-402/05 P and C-415/05 P, *Yassin Abdullah Kadi and Al Barakaat International Foundation v. Council of the European Union and Commission of the European Communities*, Judgment of 3 September 2008, [2008] ECR I-6351

2.1.3 European Court of Human Rights (ECtHR)

ECtHR, *James and others v. The United Kingdom*, Application No. 8793/79, Judgment of 21 February 1986, [1986] ECHR Ser. A No. 98
ECtHR, *Zander v. Sweden*, Application No. 14282/88, Judgment of 25 November 1993, [1993] ECHR Ser. A No. 279B
ECtHR, *Elci and others v. Turkey*, Application No. 23145/93 and 25091/94, Judgment of 13 November 2003, http://hudoc.echr.coe.int/
ECtHR, *Ostrovar v. Moldova,* Application No. 35207/03, Judgment of 13 September 2005, http://hudoc.echr.coe.int/

2.1.4 International Centre for Settlement of Investment Disputes (ICSID)

ICSID, *Southern Pacific Properties (Middle East) Limited v. Arab Republic of Egypt*, Case No. ARB/84/3, Award of 20 May 1992
ICSID, *Compañía de Aguas del Aconquija, S.A. & Compagnie Générale des Eaux v. Argentine Republic*, Case No. ARB/97/3, Award of 21 November 2000

ICSID, *Compañía de Aguas del Aconquija, S.A. & Compagnie Générale des Eaux v. Argentine Republic*, Case No. ARB/97/3, Decision on Annulment of 3 July 2002

ICSID, *Tecnicas Medioambientales Tecmed S.A. v. The United Mexican States*, Case No. ARB(AF)/00/2, Award of 29 May 2003

ICSID, *Aguas del Tunari, SA v. Republic of Bolivia*, Case No. ARB/02/3, Decision on Jurisdiction of 21 October 2005

ICSID, *Aguas Provinciales de Santa Fe S.A., Suez, Sociedad General de Aguas de Barcelona S.A. and InterAguas Servicios del Agua S.A. v. Argentine Republic*, Case No. ARB/03/17, Order in response to a petition for participation as *amicus curiae* of 17 March 2006

ICSID, *Azurix Corporation v. Argentine Republic*, Case No. ARB/01/12, Award of 14 July 2006

2.1.5 Permanent Court of International Justice (PCIJ)

PCIJ, *S.S. "Lotus"* (France v. Turkey), Judgment of 7 September 1927, PCIJ Ser. A No. 10 (1927), 28

2.1.6 Others

Trail Smelter Arbitration (United States v. Canada), Award of the US-Canadian Arbitral Tribunal of 16 April 1938 and 11 March 1941, UNRIAA Vol. III (1949), 1905

African Commission on Human and Peoples' Rights, *Free Legal Assistance Group and Others v. Zaire*, Communication No. 25/89, 47/90, 56/91, 100/93, October 1995, http://www.achpr.org/files/sessions/18th/comunications/25.89-47.90-56.91-100.93/achpr18_25.89_47.90_56.91_100.93_eng.pdf

WTO Appellate Body, *European Communities—conditions for the granting of tariff preferences to developing countries*, WT/DS246/AB/R, 7 April 2004

2.2 National Courts

2.2.1 Belgium

Belgian Court of Arbitration, *A.s.b.l. Syndicat national des propriétaires et autre*, Case No. 9/1996, 8 February 1996, in: *Moniteur belge*, 1996(02)00035, quoted after Armeni (2008), p. 4

Belgian Court of Arbitration, *Commune de Wemmel*, Case No. 36/1998, 1 April 1998, in: *Moniteur belge*, 24 April 1998, quoted after Armeni (2008), p. 5

2.2.2 France

Grand Instance Court of Avignon, *François X and the Union Fédérale des Consommateurs d'Avignon v. Société Avignonnaise des Eaux*, Case No. 1492/95, 12 May 1995, http://www.cace.fr/jurisprudence/rets/eaupression/tgi12051995.html

Tribunal de Grande Instance (District Court) of Roanne, *Revue CLCV*, No. 97, 11 March 1996

Grand Instance Court of Privas, *Compagnie de services et d'environnement v. Association des consommateurs de la Fontaulière*, Case No. 9800223, 5 March 1998, http://www.cace.fr/jurisprudence/rets/acf/tgi05031998.html

Tribunal de Grande Instance (District Court) of Meaux, *xxx. v. xxx.*, 28 February 2001, in: *Droit Monde*, No. 37–38, 2004, p. 77, quoted after COHRE (2008), p. 295

Cour d'Appel of Rennes, Decision from 2003, quoted after OHCHR, "Statement by the Permanent Mission of France to the Office of the United Nations High Commissioner on Human Rights", 2007, http://www2.ohchr.org/english/issues/water/contributions/France.pdf, p. 9

2.2.3 Germany

2.2.3.1 Federal Administrative Court of Germany (BVerwG)

BVerwG, Case No. V C 78/54, 24 June 1954, BVerwGE 1, 159
BVerwG, Case No. 1 C 136/80, 24 June 1982, BVerwGE 66, 29
BVerwG, Case No. 4 C 30/88, 17 March 1989, BVerwGE 81, 347
BVerwG, Case No. 6 C 2/97, 17 December 1997, BVerwGE 106, 64
BVerwG, Case No. 7 C 36/07, 27 September 2007, BVerwGE 129, 296

2.2.3.2 Federal Constitutional Court of Germany (BVerfG)

BVerfG, Case No. 1 BvR 220/51, 19 December 1951, BVerfGE 1, 97
BVerfG, Case No. 1 BvR 394/58, 29 July 1959, BVerfGE 10, 89
BVerfG, Case No. 1 BvR 624/56, 3 December 1969, BVerfGE 27, 253
BVerfG, Case No. 1 BvR 636/68, 4 May 1971, BVerfGE 31, 58
BVerfG, Case No. 1 BvL 4/74, 18 June 1975, BVerfGE 40, 121
BVerfG, Case No. 1 BvL 14/76, 21 June 1977, BVerfGE 45, 187
BVerfG, Case No. 1 BvL 30/76, 3 April 1979, BVerfGE 51, 115
BVerfG, Case No. 1 BvL 77/78, 15 July 1981, BVerfGE 58, 300
BVerfG, Case No. 2 BvR 458/80 and 2 BvR 486/80, 19 October 1983, BVerfGE 65, 182
BVerfG, Case No. 1 BvL 28/82, 20 March 1984, BVerfGE 66, 248

BVerfG, Case No. 1 BvL 2/82, 22 October 1985, BVerfGE 71, 66
BVerfG, Case No. 1 BvR 762/85, 20 May 1987, BVerfGE 75, 348
BVerfG, Case No. 1 BvL 20/84, 1 BvL 26/84 and 1 BvL 4/86, 29 May 1990, BVerfGE 82, 60
BVerfG, Case No. 1 BvR 1213/85, 2 March 1993, BVerfGE 88, 103
BVerfG, Case No. 1 BvL 18/93, 1 BvL 5/94, 1 BvL 6/94, 1 BvL 7/94, 1 BvR 403/94 and1 BvR 569/94, 24 January 1995, BVerfGE 92, 112
BVerfG, Case No. 2 BvR 413/88 and 2 BvR 1300/93, 7 November 1995, BVerfGE 93, 319
BVerfG, Case No. 2 BvL 5/00, 8 June 2004, BVerfGE 110, 412
BVerfG, Case No. 1 BvL 10/10 and 1 BvL 2/11, 18 July 2012, *Neue Zeitschrift Verwaltungsrecht* 31(16):1024–1033 (with comment by Tiedemann)

2.2.3.3 Federal Court of Justice of Germany (BGH)

BGH, Case No. VI ZR 349/88, 19 September 1989, BGHZ 108, 305

2.2.4 India

2.2.4.1 Allahabad High Court

Allahabad High Court, *S.K. Garg v. State of U.P.*, 28 May 1998, (1988) 2 UPLBEC 1211, http://indiankanoon.org/doc/898522/

2.2.4.2 High Court of Kerala

High Court of Kerala, *Attakoya Thangal v. Union of India*, 1 January 1990, http://www.indiankanoon.org/doc/1980528
High Court of Kerala, *F.K. Hussain v. Union of India*, 26 February 1990, AIR 1990 Ker 321, http://www.indiankanoon.org/doc/1965629

2.2.4.3 Supreme Court of India

Supreme Court of India, *Francis Coralie Mullin v. The Administartor, Union Territory of Delhi*, 13 January 1981, 1981 SCR (2) 516, http://www.indiankanoon.org/doc/78536
Supreme Court of India, *M.C. Mehta v. Union of India and others*, 12 January 1988, 1988 SCR (2) 530, http://indiankanoon.org/doc/59060
Supreme Court of India, *Subhash Kumar v. State of Bihar*, 9 January 1991, 1991 SCR (1) 5, http://indiankanoon.org/doc/1646284

Supreme Court of India, *Virendra Gaur and Others v. State of Haryana and Others*, 24 November 1994, http://indiankanoon.org/doc/114802
Supreme Court of India, *Vellore Citizens Welfare Reform v. Union of India*, 28 August 1996, http://indiankanoon.org/doc/1934103
Supreme Court of India, *Narmada Bachao Andolan v. Union of India*, 18 October 2000, http://indiankanoon.org/doc/1938608

2.2.5 South Africa

2.2.5.1 Constitutional Court of South Africa

Constitutional Court of South Africa, *The Government of the Republic of South Africa and Others v. Irene Grootboom and Others*, Case No. CCT 11/00, Judgment of 4 October 2000, 2000 (11) BCLR 1169, CC
Constitutional Court of South Africa, *Lindiwe Mazibuko and Others v. City of Johannesburg and Others*, Case No. CCT 39/09, Judgment of 8 October 2009, http://www.saflii.org/za/cases/ZACC/2009/28.pdf

2.2.5.2 High Court of South Africa

High Court (Durban and Coast Local Division) of South Africa, *Manqele v. Durban Transitional Metropolitan Council*, Judgment of 7 February 2001, 2002 6 SA 423 (D)
High Court (Witwatersrand Local Division) of South Africa, *Residents of Bon Vista Mansions v. Southern Metropolitan Local Council*, Case No. 01/12312, Judgment of 5 September 2001, 2002 (6) BCLR 625 (W)
High Court (Witwatersrand Local Division) of South Africa, *Lindiwe Mazibuko and Others v. City of Johannesburg and Others*, Case No. 13865/06, Judgment of 30 April 2008, http://www.legalbrief.co.za/filemgmt_data/files/Mazibuko%20%26%204%20Others%20v%20The%20City%20of%20Johannesburg.pdf

2.2.5.3 Supreme Court of South Africa

Supreme Court of Appeal of South Africa, *The City of Johannesburg and Others v. Lindiwe Mazibuko and Others*, Case No. 489/08, Judgment of 25 March 2009, http://www.saflii.org/za/cases/ZASCA/2009/20.pdf

2.2.6 USA

US Supreme Court, *Medellín v. Texas*, No. 06-984, 25 March 2008, 552 US 491 (2008)

3. Directives/Regulations/Resolutions of the European Union

Council of the European Communities, *Council Directive 75/440/EEC of 16 June 1975 concerning the quality required of surface water intended for the abstraction of drinking water in the Member States*, OJ 1975 L 194, 26
Council of the European Communities, *Council Directive 80/68/EEC of 17 December 1979 on the protection of groundwater against pollution caused by certain dangerous substances*, OJ 1980 L 20, 43
Council of the European Union, *Council Directive 98/83/EC of 3 November 1998 on the quality of water intended for human consumption* (European Drinking Water Directive), OJ 1998 L 330, 32
European Parliament and Council of the European Union, *Directive 2000/60/EC of the European Parliament and of the Council of 23 October 2000 establishing a framework for Community action in the field of water policy*, OJ 2000 L 327, 1
European Parliament and Council of the European Union, *Directive 2002/22/EC of the European Parliament and of the Council of 7 March 2002 on universal service and users' rights relating to electronic communications networks and services (Universal Service Directive)*, OJ 2002 L 108, 51
Council of the European Union, *Council Regulation (EC) No 881/2002 of 27 May 2002 imposing certain specific restrictive measures directed against certain persons and entities associated with Usama bin Laden, the Al-Qaeda network and the Taliban, and repealing Council Regulation (EC) No 467/2001 prohibiting the export of certain goods and services to Afghanistan, strengthening the flight ban and extending the freeze of funds and other financial resources in respect of the Taliban in Afghanistan*, No. 881/2002, 27 May 2002, OJ 2000 L 139, 9
European Parliament, *Resolution on the Commission communication on water management in developing countries and priorities for EU development cooperation*, 4 September 2003, OJ C 76 E, 430
European Parliament and Council of the European Union, *Regulation (EU) No. 211/2011 of the European Parliament and of the Council of 16 February 2011 on the citizens' initiative*, OJ 2011 L 65, 1

4. International Treaties/Conventions

Geneva Convention (III) Relative to the Treatment of Prisoners of War (1949), UNTS 75 (1950), 135
Geneva Convention (IV) Relative to the Protection of Civilian Persons in Time of War (1949), UNTS 75 (1950), 287
Council of Europe, The European Convention for the Protection of Human Rights and Fundamental Freedoms, 4 November 1950, UNTS 213 (1950)

Council of Europe, European Social Charter, 18 October 1961, CETS No. 035
Convention on the Settlement of Investment Disputes between States and Nationals of Other States (ICSID-Convention), 18 March 1965, UNTS 575 (1966), 159
International Covenant on Economic, Social and Cultural Rights (ICESCR) (1966), UNTS 993 (1976), 3
International Covenant on Civil and Political Rights (ICCPR) (1966), UNTS 999 (1976), 171
Additional Protocol (I) to the Geneva Conventions relating to the Protection of Victims of International Armed Conflicts (1977), UNTS 1125 (1979), 3
Additional Protocol (II) Relating to the Protection of Victims of Non-International Armed Conflicts (1977), UNTS 1125 (1979), 609
Vienna Convention on the Law of Treaties (VCLT), 23 May 1969, UNTS 1155 (1980), 331
Convention against Torture and Other Cruel, Inhuman or Degrading Treatment or Punishment, 10 December 1984, UNTS 1465 (1987), 85
United Nations Convention on the Law of the Sea (UNCLOS), 10 December 1982, UNTS 1833 (1994), 3
Convention on the Rights of the Child (CRC), 20 November 1989, 1577 UNTS, 3
African Charter on the Rights and Welfare of the Child (AfCRC), OAU, Doc. CAB/LEG/24.9/49 (1990)
Council of Europe, European Social Charter (revised), 3 May 1996, CETS No. 163
Charter of Fundamental Rights of the European Union (CFREU), 18 December 2000, OJ C 364/1 (2000), 1
Protocol to the African Charter on Human and People's Rights on the Rights of Women in Africa, 11 July 2003, http://www.africa-union.org/root/au/Documents/Treaties/Text/Protocol%20on%20the%20Rights%20of%20Women.pdf
Treaty on the European Union (TEU), OJ 2012 C 326, 13
Treaty on the Functioning of the European Union (TFEU), OJ 2012 C 326, 47

5. Miscellaneous Documents

European Commission, *Communication from the Commission: Services of General Interest in Europe*, COM(96) 443, 11 September 1996, OJ 1996 C 281, 3
Inter-American Commission on Human Rights, R*eport on the Situation of Human Rights in Ecuador*, OEA/Ser.L/V/II.96, Doc. 10 rev. 1, 24 April 1997, http://www.cidh.oas.org/countryrep/ecuador-eng/index%20-%20ecuador.htm
European Commission, *Communication from the Commission: Services of General Interest in Europe*, COM(2000) 580, 20 September 2000, OJ 2001 C 17, 4
Chamber of Representatives of Belgium (Commission spéciale mondialisation), *Proposition de résolution: Accès à l'eau pour chacun*, Doc. 51 1666/002, 14 March 2005

Conseil Fédéral du Développement Durable, *Avis sur la révision de la constitution concernant le droit à l'eau*, CFDD Info Bulletin No. 26, March 2006, http://www.cfdd.be/DOC/pub/info/i27f.pdf

World Water Council, *Right to Water—Moving towards a global consensus*, March 2007, http://www.worldwaterforum5.org/fileadmin/WWF5/Documents/Story_RTW_CD_March07.pdf

Amnesty International, *UN reaffirms legal rights to water and sanitation*, 2 October 2010, http://www.amnesty.org.au/news/comments/23808

European Commission, *European Citizens' Initiative hits 1 million signatures*, Press release IP/13/107, 11 February 2013

Bibliography

Akehurst M (1975) Custom as a source of international law. Br Yearb Int Law 47:1–53

Albuquerque C (2010) Chronicle of an announced birth: the coming into life of the optional protocol to the international covenant on economic, social and cultural rights – the missing piece of the international bill of human rights. Hum Rights Q 32(1):144–178

Albuquerque C, Roaf V (2012) On the right track: good practices in realising the rights to water and sanitation. Office of the High Commissioner for Human Rights. http://www.ohchr.org/Documents/Issues/Water/BookonGoodPractices_en.pdf

Albuquerque C, Winkler IT (2010) Neither friend nor foe: why the commercialisation of water and sanitation services is not the main issue for the realization of human rights. Brown J World Aff 17(1):167–179

Ali A (2002) A conceptual framework for environmental justice based on shared but differentiated responsibilities. Research Centre for Social and Economic Research on the Global Environment (University of Norfolk). Working Paper, EDM 01–02

Alston P (1984) Conjuring up new human rights: a proposal for quality control. Am J Int Law 78 (3):607–635

Alston P (1991) Revitalising United Nations work on human rights and development. Melb Univ Law Rev 18(2):216–257

Alston P (2005) Ships passing in the night: the current state of the human rights and development debate seen through the lens of the millennium development goals. Hum Rights Q 27 (3):755–829

Ambrosius G (2000) Service Public, Leistungen der Daseinsvorsorge oder Universaldienste? Zur historischen Dimension eines zukünftigen Elements europäischer Gesellschaftspolitik. In: Cox H (ed) Daseinsvorsorge und öffentliche Dienstleistungen in der Europäischen Union, 1st edn. Nomos, Baden-Baden, pp 15–43

Amor A (1998) The mandate of the UN Special Rapporteur. Emory Int Law Rev 12(2):945–950

An-Na'im A (2004) To affirm the full human rights standing of economic, social and cultural rights. In: Ghai Y, Cottrell J (eds) Economic, social and cultural rights in practice – the role of judges in implementing economic, social and cultural rights. Interrights, London, pp 7–16

Annan K (2006) Secretary-General's message to the third session of the Human Rights Council, Geneva, 29 November 2006. http://www.un.org/sg/statements/index.asp?nid=2333

Aquawal (2012) Walloon drinking water and waste-water treatment statistics: 2012 report. http://www.aquawal.be/servlet/Repository/walloon-drinking-water-and-waste-water-treatment-statistics-2012-report.pdf?ID=4304&saveFile=true

Armeni C (2008) The right to water in Belgium. International Environmental Law Research Center, Briefing Paper 2008-02

Arnold C (1978) Analyses of rights. In: Kamenka E, Tay AES (eds) Human rights. E. Arnold, Port Melbournce, pp 74–86

DOI 10.1007/978-3-642-33908-0, © Springer-Verlag Berlin Heidelberg 2014

Baderin MA, Ssenyonjo M (2010) International human rights law: six decades after the UDHR and beyond. Ashgate, Farnham

Barlow M (2007) Blue covenant: the global water crisis and the coming battle for the right to water. New Press, New York

Barlow M (2010) Access to clean water is most violated human right. http://www.theguardian.com/commentisfree/cif-green/2010/jul/21/access-clean-water-human-right

Barlow M, Clarke T (2002) Blue Gold: the fight to stop the corporate theft of the world's water. New Press, New York

Baur F (1982) Die "Naßauskiesung" – oder wohin treibt der Eigentumschutz? Neue Jurist Wochenzeitschr 35(32):1734–1736

Baxi U (2006) The future of human rights. Oxford University Press, New York

Beckett S (2006) Waiting for godot, 5th edn. Faber and Faber, London

Beisheim M (2007) Ware Wasser: Private Beteiligung bei der Wasserver- und -entsorgung in Entwicklungsländern: Lehren aus dem Fall Cochabamba. In: Rudolf B (ed) Menschenrecht Wasser? Peter Lang, Frankfurt am Main, pp 109–121

Bentham J (1843) Anarchical fallacies. In: Bauring J (ed) The works of Jeremy Bentham, vol II. W. Tait, Edinburgh

Bernsdorff N (1997) Positivierung des Umweltschutzes im Grundgesetz (Art. 20a GG). Natur Recht 19(7):328–334

Besson S (2009) European legal pluralism after Kadi. Eur Constit Law Rev 5(2):237–264

Bleisch B (2006) The human right to water – normative foundations and ethical implications. Ethics Econ 4(2). http://www.ethique-economique.net

Bluemel EB (2004) The implications of formulating a human right to water. Ecol Law Q 31 (4):957–1006

Boseley S, Goldenberg S (2005) Brazil spurns US terms for aids help. http://www.theguardian.com/world/2005/may/04/brazil.aids

Bourquain K (2008) Freshwater access from a human rights perspective: a challenge to international water law and human rights law. Martinus Nijhoff, Leiden

Boury P (2008) Tarification de l'eau en Belgique – Un accès équitable à l'eau. Paper presented at Journée Scientifique et technique ENGEES-CEMAGREF, named Les politiques d'accès à l'eau des public défavorisés mises en ouvre ici et ailleurs: regards croisés. http://www.oieau.fr/IMG/pdf/Boury.pdf

Bowett DW (1957) Estoppel before international tribunals and its relation to acquiescence. Br Yearb Int Law 33:176–202

Boyle A (2009) Polluter pays. In: Wolfrum R (ed) Max Planck encyclopedia of public international law. Oxford University Press, Oxford (online edition)

Bradley CA, Goldsmith JL (1997) Customary international law as federal common law: a critique of the modern position. Harv Law Rev 110(4):815–876

Branco MC, Henriques PD (2010) The political economy of the human right to water. Rev Radic Polit Econ 42(2):142–155

Brinkmeier S (2002) Kontrahierungszwang in der Wasserwirtschaft. Lit Publishing, Münster

Brooks RE (2005) Failed states, or the state as failure. Univ Chic Law Rev 72(4):1159–1196

Broughton J (2002) Descartes's method of doubt. Princeton University Press, Princeton

Brown W (2004) The most we can hope for...: human rights and the politics of fatalism. South Atl Q 103(2/3):451–463

Buchanan A (2004) Justice, legitimacy and self-determinant, moral foundations for international law, 1st edn. Oxford University Press, Oxford

Bueno de Mesquita J, Hunt P, Khosla R (2010) The human rights responsibility of international assistance and cooperation in health. In: Gibney M, Skogly S (eds) Universal human rights and extraterritorial obligations. University of Pennsylvania Press, Philadelphia, pp 104–129

Bullinger M (2003) Französischer service public und deutsche Daseinsvorsorge. Juristenzeitung 58(12):597–604

Búrca G (2009) The EU, the European court of justice and the international legal order after Kadi. Harv Int Law J 51(1):1–49
Busche J (1999) Privatautonomie und Kontrahierungszwang. Mohr Siebeck, Tübingen
Busse von Colbe W (2011) Entregulierung für die deutsche Wasserwirtschaft. In: Joost D (ed) Festschrift für Franz Jürgen Säcker zum 70. Geburtstag. C H Beck, Munich, pp 575–587
Byers M (1999) Custom, power, and the power of rules: international relations and customary international law. Cambridge University Press, Cambridge
Cahill A (2005) The human right to water – a right of unique status: the legal status and normative content of the right to water. Int J Hum Rights 9(3):389–410
Campbell T (2006) Rights: a critical introduction. Routledge, Abingdon
Cançado Trinidade AA (1991) The parallel evolutions of international human rights protection and of environmental protection and the absence of restrictions upon the exercise of recognized human rights. Rev Inst Interamericano Derechos Humanos 13:48–58
Castle S (2011) 18 months after vote, Belgium has government. http://www.nytimes.com/2011/12/02/world/europe/belgium-forms-government-with-elio-di-rupo-as-premier.html?_r=0
Center on Housing Rights and Evictions (COHRE) (2008) Legal resources for the right to water and sanitation: international and national standards, 2nd edn. http://www.worldwatercouncil.org/fileadmin/wwc/Programs/Right_to_Water/Pdf_doct/RWP-Legal_Res_1st_Draft_web.pdf
Centre for International Governance (2010) The role of the Special Rapporteurs. http://www.law.leeds.ac.uk/assets/files/research/cfig/special-rapporteurs-workshop-report.pdf
Charlesworth H (2008) Universal declaration of human rights (1948). In: Wolfrum R (ed) Max Planck encyclopedia of public international law. Oxford University Press, Oxford (online edition)
Charney JI (1993) Universal international law. Am J Int Law 87(4):529–551
Cheng B (1965) United Nations resolutions on outer space: "instant" international customary law? Indian J Int Law 5:23–48
Chodosh H (1991) Neither treaty nor custom: the emergence of declarative international law. Tex Int Law J 26(1):87–124
Committee on Economic, Social and Cultural Rights (ESCR-Committee) (1991) Fact Sheet No. 16 (Rev. 1). http://www.ohchr.org/Documents/Publications/FactSheet16rev.1en.pdf
Coomans F (2002) The role of the UN committee on economic, social and cultural rights in strengthening implementation and supervision of the international covenant on economic, social and cultural rights. Verfassung Recht Ü bersee 35(2):182–200
Cornut P et al (2006) Public, "club" and individual management of natural resources: the case of domestic rainwater tanks in Belgium. Die Erde Contrib Hum Geogr 137(4):273–292
Craig P, de Búrca G (2011) EU law: text, cases, and materials, 5th edn. Oxford University Press, Oxford
Cranston M (1967) Human rights – real and supposed. In: Raphael DD (ed) Political theory and the rights of men. Indiana University Press, Bloomington, pp 43–53
Cranston M (1973) What are human rights? Bodley Head, New York
Craven M (1995) The international covenant on economic, social and cultural rights – a perspective on its development. Clarendon, Oxford
Craven M (2006) Some thoughts on the emergent right to water. In: Riedel E, Rothen P (eds) The human right to water. Berliner Wissenschafts-Verlag, Berlin, pp 37–48
Crawford J (2012) Brownlie's principles of public international law, 8th edn. Oxford University Press, Oxford
Curley EM (1999) Descartes against the skeptics. Harvard University Press, Cambridge
Czychowski M, Reinhardt M (2010) Wasserhaushaltsgesetz: WHG Kommentar. C H Beck, Munich
D'Amato AA (1969) Consent, estoppel, and reasonableness: three challenges to Universal International Law. Va J Int Law 10(1):1–31
D'Amato AA (1971) The concept of custom in international law. Cornell University Press, Ithaca
Dembour MB (2006) Who believes in human rights? Reflections on the European convention. Cambridge University Press, Cambridge

Dembour MB (2010) What are human rights? Four schools of thought. Hum Rights Q 32(1):1–20

Dennis MJ, Stewart DP (2004) Justiciability of economic, social and cultural rights: should there be an international complaints mechanism to adjudicate the rights to food, water, housing and health? Am J Int Law 98(3):462–515

Department for International Development (DFID) (2001) Addressing the water crisis: healthier and more productive lives for poor people. http://webarchive.nationalarchives.gov.uk/+/http://www.dfid.gov.uk/pubs/files/tspwater.pdf

Department of Water Affairs (2010) Annual Report 2009/2010 (Part 1). http://www.dwaf.gov.za/documents/AnnualReports/2010AnnualReport.pdf

Department of Water Affairs (2012) Annual report 2011/2012. http://www.dwaf.gov.za/documents/AnnualReports/DWA%20AR%202011-12%20Reduced.pdf

Department of Water Affairs and Forestry (2008) Annual report 2007/2008. http://www.dwaf.gov.za/documents/AnnualReports/ANNUALREPORT2007-2008.pdf

Descartes R (1999) Discourse on method and meditations on first philosophy (trans: Cress DA), 4th edn. Hackett Publishing Company, Indianapolis

Dinstein Y (1981) The right to life, physical integrity and liberty. In: Henkin L (ed) The international bill of rights. Columbia University Press, New York, pp 114–120

Donnelly J (2003) Universal human rights in theory and practice, 2nd edn. Cornell University Press, Ithaca

Dreyer J, Bartl U (2010) Preishöhenkontrolle bei Waffenlieferungen – Wasserpreise Wetzlar. Neue Jurist Wochenschr 63(35):2553–2554

Dugger CW (2008) Cholera epidemic sweeping across crumbling Zimbabwe. http://www.nytimes.com/2008/12/12/world/africa/12cholera.html?_r=0

Dupuy PM (2006) Le droit à l'eau, un droit international? La mise en oeuvre du droit à l'eau, Institut International de droit d'expression et d' Inspiration Francaises, Actes du XXIXe Congrès ordinaire de l'IDEF, 2006, 29 September – 1st October 2005; at the same European University Institute Working Paper Law 2006/06, Florence

Dupuy PM (2009) Unification rather than fragmentation of international law? The case of international investment law and human rights law. In: Dupuy PM, Francioni F, Petersmann EU (eds) Human rights in international investment law and arbitration. Oxford University Press, Oxford, pp 45–62

Dworkin R (1984) Rights as trumps. In: Waldron J (ed) Theories of rights. Oxford University Press, Oxford, pp 153–167

Eckes C (2008) Judicial review of European anti-terrorism measures – the *Yusuf* and *Kadi* judgments of the court of first instance. Eur Law J 14(1):74–92

Emmerich-Fritsche A (2007) Privatisierung der Wasserversorgung in Bayern und kommunale Aufgabenverantwortung. Bayrische Verwaltungsblätter 138(1):1–8

European Committee of Social Rights (ECSR) (2006) Digest of the Case Law. http://www.sm.ee/fileadmin/meedia/Dokumendid/Sotsiaalvaldkond/sotsiaalharta/Case_law_Digest_en_1__1_.pdf

European Committee of Social Rights (ECSR) (2012) Activity report 2011. http://www.coe.int/t/dghl/monitoring/socialcharter/Presentation/ActivityReport2011_en.pdf

Faßbender K (2001) Gemeinschaftsrechtliche Anforderungen an die normative Umsetzung der neuen EG-Wasserrahmenrichtlinie. Neue Zeitschrift Verwaltungsrecht 20(3):241–249

Federal Ministry of the Environment, Nature Conservation and Nuclear Safety (BMU) (2011) Wasser – Wohlstand – Wandel: Gewässerschutz sichert unsere Grundlagen für Leben, Vielfalt und Nachhaltigkeit. http://www.bmu.de/fileadmin/bmu-import/files/pdfs/allgemein/application/pdf/broschuere_wasser_wohlstand_wandel_bf.pdf

Fidler D (1996) Challenging the classical concept of custom: perspectives on the future of customary international law. Ger Yearb Int Law 39:198–248

Filmer-Wilson E (2005) The human rights-based approach to development: the right to water. Neth Q Hum Rights 23(2):213–241

Fischer M, Zwetkow K (2003) Privatisierungsoptionen für den deutschen Wasserversorgungsmarkt. Zeitschrift Wasserrecht 42(3):129–156

Ford S (2010) Is the failure to respond appropriately to a natural disaster a crime against humanity? The responsibility to protect and individual criminal responsibility in the aftermath of cyclone Nargis. Denver J Int Law Policy 38(2):227–276
Forsthoff E (1938) Die Verwaltung als Leistungsträger. Kohlhammer, Stuttgart
Forsthoff E (1959) Rechtsfragen der leistenden Verwaltung. Kohlhammer, Stuttgart
Francis R (2005) Water justice in South Africa: natural resources policy at the intersection of human rights, economics, and political power. Georget Int Environ Law Rev 18(1):149–196
Franklin B (1793) Poor Richard's Almanack. Quoted in: McNeill JR (2001) Something new under the sun: an environmental history of the twentieth-century world. W W Norton, New York
Friauf KH (1971) Zur Rolle der Grundrechte im Interventions- und Leistungsstaat. Deutsches Verwaltungsblatt 86(18):674–682
Furch L (2010) Menschenrecht auf Wasser- und Sanitärversorgung: UN-Resolutionen als Schlüssel zum Paradies? Eur J Transnatl Stud 2(2):26–49
Gassner E (2011) Rechtskonkretisierung zum Schutz der natürlichen Lebensgrundlagen nach Art. 20a GG. Natur Recht 33(5):320–323
Geiger R (2010) Commentary on article 11 EUT. In: Geiger R, Khan DE, Kotzur M (eds) EUV/AEUV Kommentar, 5th edn. C H Beck, Munich
German Federal Foreign Office (AA) (2010) Germany in the United Nations. https://www.auswaertiges-amt.de/cae/servlet/contentblob/382774/publicationFile/4262/DeutschlandindenVN-englisch.pdf
Gleick PH (1998) The human right to water. Water Policy 1(5):487–503
Gleick PH (2010) The human right to water (and sanitation). http://www.huffingtonpost.com/peter-h-gleick/the-human-right-to-water_b_671175.html
Gleick PH, Wolf G, Calecki EL et al (2002) The new economy of water: the risks and benefits of globalization and privatization of freshwater. Report by the Pacific Institute for Studies in Development, Environment and Security, February 2002
Glennon R (2005) Water scarcity, marketing and privatization. Tex Law Rev 83(7):1873–1902
Goeters H (2008) Das institutionelle Gleichgewicht – seine Funktion und Ausgestaltung im Europäischen Gemeinschaftsrecht. Duncker & Humblot, Berlin
Goldsmith QC (2001) A charter of rights, freedoms and principles. Common Mark Law Rev 38 (5):1201–1216
Gorbachev M (2010) The right to water. http://www.nytimes.com/2010/07/17/opinion/17iht-edgorbachev.html?_r=0
Graham P (1996) The will theory of rights: a defense. Law Philos 15(3):257–270
Grimes HJ (2009) Responding to the water crisis: the complementary roles of water governance and the human right to water. J Water Law 20(2–3):119–128 (special issue: Promoting water for all)
Grotius H (2009) The Freedom of the Seas or: the right which belongs to the Dutch to take part in the East Indian Trade (transl: Ralph van DemanMagoffin), 1608 ed version. Oxford University Press, New York
Habermas J (1996) Between facts and norms: contributions to a discourse theory of law and democracy. MIT Press, Cambridge/MA
Halberstam D, Stein E (2009) The United Nations, the European Union and the king of Sweden: economic sanctions and individual rights in a plural world order. Common Mark Law Rev 46 (1):13–72
Hall D, Lobina E (2007) International actors and multinational water company strategies in Europe, 1990–2003. Utilities Policy 15(2):64–77
Hall D, Lobina E (2012) Water companies and trends in Europe 2012. Public Services International Research Unit. http://www.epsu.org/IMG/pdf/2012_Water_companies-EWCS.pdf
Hall D, Lobina E, de la Motte R (2004) Making water privatisation illegal: new laws in Netherlands and Uruguay. http://gala.gre.ac.uk/3769/1/PSIRU_9343_-_2004-11-W-crim.pdf
Hall D, Lobina E, de la Motte R (2005) Public resistance to privatization in water and energy. Dev Pract 15(3–4):286–301

Hannum H (1995–1996) The status of the universal declaration of human rights in national and international law. Ga J Int Comp Law 25(1–2):287–398

Hannum H (2007) Reforming the special procedures and mechanisms of the commission on human rights. Hum Rights Law Rev 7(1):73–92

Hardin G (1968) The tragedy of the commons. Science 162(3859):1243–1248

Harris PG (1999) Common but differentiated responsibility. N Y Univ Environ Law J 7(1):27–48

Harris PG (2006) Bechtel, Bolivia resolve dispute/company drops demand over water contract canceling. http://www.sfgate.com/news/article/Bechtel-Bolivia-resolve-dispute-Company-drops-2523974.php

Hart HLA (1954) Definition and theory in jurisprudence. Law Q Rev 70(1):37–60

Hart HLA (1973) Bentham on legal rights. In: Simpsons AWB (ed) Oxford essays in jurisprudence. Second series. Clarendon, Oxford, pp 171–201

Heintschel von Heinegg W (2004) Kapitel 4: Die weiteren Quellen des Völkerrechts. In: Ipsen K (ed) Völkerrecht, 5th edn. C H Beck, Munich, pp 210–256

Henneke HG (2011) Kommentierung von Art. 28. In: Schmidt-Bleibtreu B, Hofmann H, Hopfauf A (eds) Kommentar zum Grundgesetz, 12th edn. Heymann, Cologne

Herdegen M (2012) Commentary on Art. 1 Abs. 1. In: Maunz T, Dürig G (eds) Grundgesetz: Kommentar. C H Beck, Munich

Herzog R, Grzeszick B (2012) Commentary on Art. 20. In: Maunz T, Dürig G (eds) Grundgesetz: Kommentar. C H Beck, Munich

Hey E (2011) Common but differentiated responsibilities. In: Wolfrum R (ed) Max Planck encyclopedia of public international law. Oxford University Press, Oxford (online edition)

Hill TE Jr (2006) Kantian normative ethics. In: Copp D (ed) The Oxford handbook of ethical theory. Oxford University Press, Oxford, pp 480–514

Hirsch M (2009) Investment tribunals and human rights: divergent paths. In: Dupuy PM, Francioni F, Petersmann EU (eds) Human rights in international investment law and arbitration. Oxford University Press, Oxford, pp 97–115

Human C (2006) The human right to water in Africa: the South African example. In: Riedel E, Rothen P (eds) The human right to water. Berliner Wissenschafts-Verlag, Berlin, pp 83–93

Human Rights Council (HRC) (2008) Manual of operations of the special procedures of the Human Rights Council. http://www.ohchr.org/Documents/HRBodies/SP/Manual_Operations2008.pdf

Humphrey J (1979) The universal declaration of human rights: its history, impact and juridical character. In: Ramcharan BG (ed) Human rights: thirty years after the universal declaration. Springer/Kluwer, Dordrecht, pp 21–37

Ignatieff M (2001) Human rights as politics and idolatry. Pinceton University Press, Princeton

International Institute for Democracy and Electoral Assistance (IDEA) (2012) Compulsory voting – what is compulsory voting? http://www.idea.int/vt/compulsory_voting.cfm

Irmscher R (2001) Kommunale Wasserwirtschaft in einem sich ändernden Umfeld. In: Büscher E (ed) Wasserwirtschaft im Aufbruch. Luchterhand, Cologne

Isiksel NT (2010) Fundamental rights in the EU after Kadi and Al Barakaat. Eur Law J 16 (5):551–577

Jaag T (2012) Demokratische Legitimation der EU-Außenpolitik nach Lissabon. Europarecht 47 (3):309–322

Jarass HD, Pieroth B (2012) Grundgesetz für die Bundesrepublik Deutschland GG, 12th edn. C H Beck, Munich

Jeffords C, Shah F (2012) On the natural and economic difficulties to fulfilling the human right to water within a neoclassical economics framework. http://www.icape.org/h3-jeffords-shah.pdf

Jenkins MW, Sugden S (2006) Rethinking sanitation: lessons and innovation for sustainability and success in the new millennium. UNDP Human Development Report Office, New York

Kadelbach S, Kleinlein T (2006) Überstaatliches Verfassungsrecht: zur Konstitutionalisierung im Völkerrecht. Arch Vö lkerrechts 44(3):235–266

Kämmerer JA (2002) Daseinsvorsorge als Gemeinschaftsziel oder: Europas “soziales Gewissen”. Neue Zeitschrift Verwaltungsrecht 21(9):1041–1046

Kämmerer JA (2004) Strategien zur Daseinsvorsorge: Dienste im allgemeinen Interesse nach der “Altmark”-Entscheidung des EuGH. Neue Zeitschrift Verwaltungsrecht 23(1):28–34

Kelly JP (2000) The Twilight of customary international law. Va J Int Law 40(2):449–544

Kelsen H (1958) Die Einheit von Völkerrecht und staatlichem Recht. Zeitschrift ausländisches öffentliches Recht Völkerrecht 19:234–248

Kersten J (2005) Die Entwicklung des Konzepts der Daseinsvorsorge im Werk von Ernst Forsthoff. Der Staat 44(4):543–569

Kessler S (2007) Wer hat Teil am Blauen Gold? Bemerkungen zur menschenrechtlichen Dimension des Rechts aufWasser? In: Rudolf B (ed) Menschenrecht Wasser? Peter Lang, Frankfurt am Main, pp 185–194

Kirschner AJ (2011) The human right to water and sanitation. Max Planck Yearb UN Law 15:445–487

Klee K (2000) Die progressive Verwirklichung wirtschaftlicher und sozialer und kultureller Rechte. Eine Interpretation von Art. 2 Abs. 1 des Internationalen Pakts für wirtschaftliche, soziale und kulturelle Rechte. Boorberg, Stuttgart

Knauber R (1984) Die jüngere Entschädigungsrechtsprechung des BGH nach dem Naßauskiesungsbeschluß des BVerfG. Neue Zeitschrift Verwaltungsrecht 3(12):753–759

Knoll-Tudor I (2009) The fair and equitable treatment standard and human rights norms. In: Dupuy PM, Francioni F, Petersmann EU (eds) Human rights in international investment law and arbitration. Oxford University Press, Oxford, pp 310–343

Knopp GM (2010) Das neue Wasserhaushaltsrecht. C H Beck, Munich

Köck W (2012) Wasserwirtschaft und Gewässerschutz in Deutschland, Rechtsrahmen, Institutionen, Organisation. Zeitschrift Umweltrecht 23(3):140–150

Kok A (2005) Privatisation and the right to access to water. In: De Feyter K, Isa FG (eds) Privatisation and human rights in the age of globalisation. Intensentia, Antwerp, pp 259–287

Ladwig B (2007) Kann es einMenschenrecht aufWasser geben? In: Rudolf B (ed) Menschenrecht Wasser? Peter Lang, Frankfurt am Main, pp 45–58

Langford M (2006) Ambition that overleaps itself? A response to Stephen Tully’s critique of the general comment on the right to water. Neth Q Hum Rights 24(3):433–459

Langford M (2006) Expectation of plenty: response to Stephen Tully. Neth Q Hum Rights 24 (3):473–479

Laskowski SR (2010) Das Menschenrecht auf Wasser: Die rechtlichen Vorgaben zur Sicherung der Grundversorgung mit Wasser und Sanitärleistungen im Rahmen einer ökologisch-nachhaltigen Wasserwirtschaftsordnung. Mohr Siebeck, Tübingen

Le Nestour C, Zinow BM (1994) Rechtsfragen des “Service Public”. Recht Energiewirtschaft 55 (4):129–133 and 55(5):170–174

Leckie S, Gallagher A (2006) Economic, social, and cultural rights: a legal resource guide. University of Pennsylvania Press, Philadelphia

Lever P (2007) Wasser: Die Global Players und ihre Strategien – Privatisierung als Fluch oder Segen? In: Rudolf B (ed) Menschenrecht Wasser. Peter Lang, Frankfurt am Main, pp 85–92

Liebenberg S (1998) The national water bill – breathing life into the right to water. Econ Soc Rights Rev 1(1):3–6

Lobina E (2000) Cochabamba: water war. Public Services International Research Unit (PSIRU). Working Paper Series at Business School of University of Greenwich

Locke J (1821) Two treatises on government: book II. Whitmore and Fenn, London

MacCormick N (1976) Children’s rights: a test-case for theories of rights. Arch Rechts Sozialphilosophie 62(3):306–316

MacFarlane LJ (1985) The theory and practice of human rights. Dartmouth Publishing Co. Ltd, Aldershot

MacGibbon IC (1957) Customary international law and acquiescence. Br Yearb Int Law 33:115–145

MacGibbon IC (1958) Estoppel in international law. Int Comp Law Q 7(3):468–513
Mamlyuk BN (2009) Analyzing the polluter pays principle through law and economics. Southeast Environ Law J 18(1):39–80
Manda LO (2003) Controversy rages over "GM" food aid. Afr Renewal 16(4):5
Marks S (2004) The human right to development: between rhetoric and reality. Harv Hum Rights J 17:137–168
Marong A (2010) Development, right to, international protection. In: Wolfrum R (ed) Max Planck encyclopedia of public international law. Oxford University Press, Oxford (online edition)
Mayeda G (2008) International investment agreements between developed and developing countries: dancing with the devil? A case commentary on the Vivendi, Sempra and Enron awards. McGill Int J Sustain Dev Law Policy 4(2):189–230
McCaffrey SC (1992) A human right to water: domestic and international implications. Georget Int Environ Law Rev 5(1):1–24
McCaffrey SC (2005) The human right to water. In: Bernasconi-Osterwalder N, Brown Weiss E, de Boisson Chazournes L (eds) Fresh water and international economic law. Oxford University Press, Oxford, pp 93–115
McDonald DA (2002) The theory and practice of cost recovery in South Africa. In: McDonald DA, Pape J (eds) Cost recovery and the crisis of service delivery in South Africa. HSRC Publishers/ZED Books, Cape Town/London, pp 17–37
McFarland Sánchez-Moreno M, Higgins T (2004) No recourse: transnational corporations and the protection of economic, social and cultural rights in Bolivia. Fordham Int Law J 27 (5):1663–1805
McIntyre A (2007) After virtue: a study in moral theory, 3rd edn. University of Notre Dame Press, Notre Dame
McNeil DG Jr (2010) Cholera outbreak kills 150 in Haiti. http://www.nytimes.com/2010/10/23/world/americas/23cholera.html
Mehta L (2005) Citizenship and the right to water. Lessons from South Africa's free basic water policy. In: Kabeer N (ed) Inclusive citizenship: meanings and expressions. Zed Books, London, pp 235–250
Mestmäcker EJ (1998) Daseinsvorsorge und Universaldienst im europäischen Kontext: Ein Beitrag zur Funktion von Art. 90 Abs. 2 EGV. In: Ruland F, von Baron Maydell B, Papier HJ (eds) Verfassung, Theorie und Praxis des Sozialstaats. Festschrift für Hans F. Zacher zum 70. Geburtstag. C F Müller, Heidelberg, pp 635–651
Miquel MG (2003) Report to the Senate No. 215 (2002–2003): La qualité de l'eau et l'assainissement en France. http://www.senat.fr/rap/l02-215-1/l02-215-1.html
Mistry R (2010) Water becomes a legal human right. http://civicusalpha.org/ehhr3.0/component/content/article/9-human-rights-news/108-water-becomes-a-legal-human-right
Möllers C (2005) Gewaltengliederung. Legitimation und Dogmatik im nationalen und internationalen Rechtsvergleich. Mohr Siebeck, Tübingen
Müller T (2011) Belgien – handlungsunfähig in der Krise. http://www.zeit.de/wirtschaft/2011-11/belgien-euro-krise
Müller JP, Cottier T (1995) Estoppel. In: Bernhardt R (ed) Encyclopedia of public international law, vol II. Elsevier North Holland, Amsterdam, pp 116–119
Muralidhar S (2004) Economic, social & cultural rights: an Indian response to the justiciability debate. In: Ghai Y, Cottrell J (eds) Economic, social & cultural rights in practice: the role of judges in implementing economic, social & cultural rights. Interrights, London, pp 23–32
Muralidhar S (2006) The right to water: an overview of the Indian legal regime. In: Riedel E, Rothen P (eds) The human right to water. Berliner Wissenschafts-Verlag, Berlin, pp 65–81
Murty MN, Kumar S (2011) Water pollution in India: an economic appraisal. In: Infrastructure Development Finance Company (ed) India infrastructure report 2011 – water: policy and performance for sustainable development. Oxford University Press, New Delhi, pp 285–298
Musil A, Kirchner S (2012) Das Recht der Berliner Verwaltung, 3rd edn. Springer, Berlin

Mydans S (2008) Myanmar faces pressure to allow major aid effort. http://www.nytimes.com/2008/05/08/world/asia/08myanmar.html?_r=1&hp
Naegele J (2004) What is wrong with full-fledged water privatization? J Law Soc Chall 6:99–130
Nettesheim M (2012) Commentary on Art 11 EUT. In: Grabitz E, Hilf M, Nettesheim M (eds) Das Recht der Europäischen Union. C H Beck, Munich
Norris AL, Metzidakis KE (2010) Public protests, private contracts: confidentiality in ICSID arbitration and the Cochabamba water war. Harv Negot Law Rev 15(1):31–76
O'Neill O (1980) Kantian approaches to some famine problems. In: Regan T (ed) Matters of life and death. Random House, New York, pp 285–294
O'Neill O (1996) Towards justice and virtue: a constructive account of practical reasoning. Cambridge University Press, Cambridge
Office of the High Commissioner for Human Rights (OHCHR) (2001) Fact sheet no. 27: seventeen frequently asked questions about United Nations Special Rapporteurs. http://www.ohchr.org/Documents/Publications/FactSheet27en.pdf
Office of the High Commissioner for Human Rights (OHCHR) (2005) Economic, social and cultural rights: handbook for national human rights institutions. United Nations, New York
Office of the High Commissioner for Human Rights (OHCHR) (2007) Study on Human Rights and the access to water: contribution of the Government of Belgium. http://www2.ohchr.org/english/issues/water/contributions/Belgium.pdf
Organisation for Economic Co-operation and Development (OECD) (2006) Liberalisation and universal access to basic services. OECD, Paris
Organisation for Economic Co-operation and Development (OECD) (2007) Environmental performance reviews: Belgium. http://www.oecd.org/env/country-reviews/environmental performancereviewsbelgium2007.htm
Ouvry B (2007) Access to water as a human right, different ways to fulfill this vital commitment: the cases of Belgium and its regions: speaking note on the occasion of the consultation on human rights and access to safe drinking water and sanitation. Office of the High Commissioner for Human Rights, Geneva, 11 May 2007. http://www.unece.org/env/water/whmop1/roundtable/Ouvry.pdf
Pahl-Wostl C, Gupta J, Petry D (2008) Governance and the global water system: a theoretical exploration. Glob Gov 14(4):419–436
Palombella G (2007) From human rights to fundamental rights. Consequences and conceptual distinction. Arch Rechts Sozialphilosophie 93(3):396–426
Pant R (2003) From communities' hands toMNCs' BOOTs: a case study from India on the right to water. http://www.righttowater.info/wp-content/uploads/india_cs.pdf
Papier HJ (2012) Commentary on Article 14. In: Maunz T, Dürig G et al (eds) Grundgesetz-Kommentar, Band I. C H Beck, Munich
Parienté J (2012) TVA de droite, TVA de gauche: le jeu des différences selon Hollande. http://www.lemonde.fr/politique/article/2012/11/14/tva-de-droite-tva-de-gauche-le-jeu-des-differences-selon-hollande_1790190_823448.html
Partzsch L (2006) Wasser in der Krise – Das Beispiel Südafrika. Solidarische Welt 196(4):4–6
Perin M (2000) Azurix water bugs Argentina. http://houston.bizjournals.com/houston/stories/2000/05/08/tidbits.html
Perry MJ (1998) The idea of human rights: four inquiries. Oxford University Press, New York
Petersmann EU (2008) De-fragmentation of international economic law through constitutional interpretation and adjudication with due respect for reasonable disagreement. Loyola Univ Chic Int Law Rev 6(1):209–248
Petersmann EU (2009) Constitutional theories of international economic adjudication and investor-state arbitration. In: Dupuy PM, Francioni F, Petersmann EU (eds) Human rights in international investment law and arbitration. Oxford University Press, Oxford, pp 138–194
Petrova V (2006) At the frontiers of the rush for Blue Gold: water privatization and the human right to water. Brooklyn J Int Law 31(2):577–613

Pindyck RS, Rubinfeld DL (2009) Microeconomics, 7th edn. Pearson Education, Upper Saddle River

Pinheiro PS (2003) Global insights: musings of a UN Special Rapporteur. Glob Gov 9(7):7–13

Pogge TW (2002) World poverty and human rights. Polity Press/Blackwell Publishing Ltd, Oxford

Powlowski M (2006) Making public health motivated evictions consistent with the right to housing. Quinnipiac Health Law J 9(2):271–318

Przetacznik F (1976) The right to life as a basic human right. Rev Droits l'Homme 9(4):585–608

Püttner G (2002) Das grundlegende Konzept der Daseinsvorsorge, Kommunale Daseinsvorsorge: Begriff, Geschichte, Inhalte. In: Hrbek R, Nettesheim M (eds) Europäische Union und mitgliedsstaatliche Daseinsvorsorge. Nomos, Baden-Baden, pp 32–38

Rawls J (1958) Justice as fairness. Philos Rev 67(2):164–194

Rawls J (1971) A theory of justice, 1st edn. Harvard University Press, Cambridge

Raz J (1984) On the nature of rights. Mind 93(370):194–214

Reed L et al (2011) Guide to ICSID arbitration, 2nd edn. Kluwer Law International, Alphen aan den Rijn

Riedel E (1986) Theorie der Menschenrechtsstandards. Duncker & Humblot, Berlin

Riedel E (1991) Standards and sources – farewell to the exclusivity of the sources triad in international law? Eur J Int Law 2(2):58–84

Riedel E (2005) The human right to water. In: Dicke K et al (eds) Weltinnenrecht – Liber amicorum Jost Delbrück. Duncker & Humblot, Berlin, pp 586–606

Riedel E (2006) The human right to water and general comment No. 15 of the CESCR. In: Riedel E, Rothen P (eds) The human right to water. Berliner Wissenschafts-Verlag, Berlin, pp 19–36

Riedel E (2011) International covenant on economic, social and cultural rights (1966). In: Wolfrum R (ed) Max Planck encyclopedia of public international law. Oxford University Press, Oxford (online edition)

Ringwald R (2008) Daseinsvorsorge als Rechtsbegriff: Forsthoff, Grundgesetz und Grundversorgung. Peter Lang, Frankfurt am Main

Risse M (2008) Immigration and common ownership of the Earth. In: Conference paper for United Nations Development Programme, Cambridge, December 2008

Risse M (2008) Original ownership of the earth: a contemporary approach. Harvard Kennedy School Working Paper Series, Cambridge

Risse M (2012) On global justice. Princeton University Press, Princeton

Risse M (2013) The human right to water and common ownership of the Earth, HKS Faculty Research Working Paper Series, RWP13-003, February 2013

Roberts A (2001) Traditional and modern approaches to customary international law: a reconciliation. Am J Int Law 95(4):757–791

Robinson N (1958) The universal declaration of human rights: its origin, significance, application, and interpretation. Institute of Jewish Affairs, World Jewish Congress, New York

Rosencranz A, Rustomjee S (1995) Citizens' right to a healthful environment in India. Environ Policy Law 25(6):324–328

Rother L (2005) Prostitution puts U.S. and Brazil at odds on AIDS policy. http://www.nytimes.com/2005/07/24/international/americas/24brazil.html

Rudolf B (2007) Menschenrecht Wasser: Herleitung, Inhalt, Bedeutung, Probleme. In: Rudolf B (ed) Menschenrecht Wasser? Peter Lang, Frankfurt am Main, pp 15–43

Rüfner W (2006) Daseinsvorsorge und soziale Sicherheit. In: Isensee J, Kirchhof P (eds) Handbuch des Staatsrechts der Bundesrepublik Deutschland, vol IV, 3rd edn, Aufgaben des Staates. C F Müller, Heidelberg

Ruge R (2003) Das Grünbuch der EG Kommission zu Leistungen der Daseinsvorsorge: Neue Konturen für einen alten Irrwisch? Zeitschrift Rechtspolitik 36(10):353–357

Santiso C (2001) Good governance and aid effectiveness: the World Bank and conditionality. Georget Public Policy Rev 7(1):1–22

Santos C, Villareal A (2006) Uruguay: direct democracy in defense of the right to water (revised version). http://www.tni.org/sites/www.tni.org/archives/books/wateruruguayrev.pdf
Scanlon J, Cessar A, Nemes N (2004) Water as a human right? International Union for Conservation of Nature and Natural Resources (IUCN) Environmental Policy and Law Paper No. 51. Gland/Cambridge
Schachter O (1991) International law in theory and practice. Nijhoff, Dordrecht
Scheinin M (2009) Is the ECJ ruling in Kadi incompatible with international law? In: Eeckhout P, Tridimas T (eds) Yearbook of European law, vol 28, pp 637–653
Scheinin M (2009) Towards a world court of human rights. Research report to the Swiss initiative to commemorate the 60th anniversary of the Universal Declaration of Human Rights. http://www.udhr60.ch/report/hrCourt_scheinin0609.pdf
Schill S (2006) Fair and equitable treatment under investment treaties as an embodiment of the rule of law. Transnatl Dispute Manag 3(5):9–30
Schwarzenberger G (1947) The inductive approach to international law. Harv Law Rev 60 (4):539–570
Schwarzenberger G (1965) The inductive approach to international law. Stevens, London
Scott C, Alston P (2000) Adjudicating constitutional priorities in a transnational context: a comment on Soobramoney's legacy and Grootboom's promise. South Afr J Hum Rights 16 (2):206–268
Seewald O (2006) Kommunalrecht. In: Steiner U (ed) Besonderes Verwaltungsrecht, 8th edn. C F Müller, Heidelberg, pp 1–170
Selden J (1636) Mare Clausum Seu De Dominio Maris. Maire, Lugduni Batavorum
Sen A (1987) On ethics and economy. Blackwell, Oxford
Sen A (2004) Elements of a theory of human rights. Philos Public Aff 32(4):315–356
Sepúlveda M (2003) The nature of the obligations under the international covenant on economic, social and cultural rights. Intersentia, Antwerp
Sepúlveda M (2006) Obligations of international assistance and cooperation in an optional protocol to the international covenant on economic, social and cultural rights. Neth Q Hum Rights 24(2):7–56
Shah P (1997) International human rights: a perspective from India. Fordham Int Law J 21 (1):24–44
Shaw W (2006) The consequentialist perspective. In: Dreier J (ed) Contemporary debates in moral theory. Wiley-Blackwell, Malden, pp 5–20
Shaw MN (2008) International law, 6th edn. Cambridge University Press, Cambridge
Shaw A, Gotora G (2013) Zimbabweans vote in favor of new constitution – Zimbabwe: 94 percent vote for new constitution, but police jail rights lawyer for 3rd night. http://readingeagle.com/article.aspx?id=459940
Shelton D (2002) Human rights and the environment: jurisprudence of human rights bodies. Environ Policy Law 32(3–4):158–167
Shestack JJ (1998) The philosophic foundations of human rights. Hum Rights Q 20(2):201–234
Shue H (1980) Basic rights: subsistence, affluence, and U.S. Foreign Policy. Princeton University Press, Princeton
Siemaszko C (2008) Myanmar government refuses U.S. aid. http://www.nydailynews.com/news/world/myanmar-government-refuses-u-s-aid-article-1.328655
Simma B, Alston P (1988–1989) The sources of human rights law: custom, jus cogens, and general principles. Aust Yearb Int Law 12:82–108
Singer P (1972) Famine, affluence, and morality. Philos Public Aff 1(1):229–243
Singer P (2004) One world: the ethics of globalization, 2nd edn. Yale Universty Press, New Haven
Skogly SI (2006) Beyond national borders: states' human rights obligations in international cooperation. Intersentia, Antwerp
Smets H (2005) La solidarité pour l'eau potable. Research Paper Series of the Academie de l'eau, Paris. http://www.academie-eau.org/admin/fichier_publication/70-2005_Pour_un_droit_effectif_a_l_eau_potable_Avril_2005-1271951606fichier_publication0.pdf

Smets H (2006) The right to water at the 4th world water forum in Mexico. http://www.righttowater.info/wp-content/uploads/The-Right-to-Water-at-the-World-Water-Forum-Mexico-2006.pdf

Smets H (2006) The right to water in national legislations. Research paper series of the Agence Française de Développement, Paris. http://www.emwis.net/documents/database/semide/PDF/right-to-water_EN

Smets H (2008) De l'eau potable a un prix abordable: La pratique des États. http://www.academie-eau.org/admin/fichier_article/65-2008_De_l_eau_potable_a_un_prix_abordable-1269361608 fichier00.pdf

Smets H (2008) Rights and duties associated with the right to water. In: Fischer-Lescano A et al (eds) Frieden und Freiheit, Festschrift für Michael Bothe zum 70. Geburtstag. Nomos, Baden-Baden, pp 711–750

Smets H (2012) Charging the poor for drinking water: the experience of continental European countries concerning the supply of drinking water to poor users. Paper presented at a workshop hosted by publicpolicy, 19 June 2012. http://www.publicpolicy.ie/wp-content/uploads/Water-for-Poor-People-Lessons-from-France-Belgium.pdf

Smith A (1790) The theory of moral sentiments, 6th edn. A. Miller, London

Solanes M, Jouravlev A (2006) Water governance for development and sustainability. CEPAL – Serie Recursos naturals e infraestructura No. 111. United Nations, Santiago

Sontag D (2012) In Haiti, global failures on a cholera epidemic. http://www.nytimes.com/2012/04/01/world/americas/haitis-cholera-outraced-the-experts-and-tainted-the-un.html

Speed S (2008) Rights in rebellion: indigenous struggle and human rights in Chiapas. Stanford University Press, Stanford

Stammers N (2009) Human rights and social movements. Pluto, London

Stein T (1988) Remarks on custom. In: Cassese A, Weiler JHH (eds) Change and stability in international law-making. Walter de Gruyter, Berlin, pp 12–14

Steiner HJ, Alston P, Goodman R (2008) International human rights in context, law, politics, morals – text and materials, 3rd edn. Oxford University Press, New York

Szabo A (2010) The value of free water: analyzing South Africa's free basic water policy. http://www.uh.edu/~aszabo2/water_szabo_0512.pdf

Taylor PL, Asworth DN, Kitchen DE (2007) Helping Africa achieve millennium development goal 1: a hyperbole-free US genetically modified food aid policy. Afr Policy J, Harvard Kennedy School of Government, 03/2007

Thielbörger P (2009) The human right to water versus investor rights: double-dilemma or pseudo-conflict? In: Dupuy PM, Francioni F, Petersmann EU (eds) Human rights in international investment law and arbitration. Oxford University Press, Oxford, pp 487–511

Thielbörger P (2013) Governing international watercourses: implications of the human right to water, chapter 3. In: Kibaroglu A, Kirschner AJ, Mehring S, Wolfrum R (eds) Water law and cooperation in the Euphrates Tigris region: a comparative and inter-disciplinary study of international and national water law. Brill, Leiden

Thielbörger P (2013) The right to water and its effective multi-level protection. In: Cremona M, Hilpold P, Lavranos N et al (eds) Reflections on the constitutionalization of international economic law – Liber amicorum Ernst-Ulrich Petersmann. Martinus Nijhoff, Leiden, pp 553–569

Thielbörger P (2013) Wasser und Würde. In: Kapust A, Gröschner R, Lembcke OW (eds) Wörterbuch der Würde. UTB Fink, Paderborn, pp 323–324

Thürer D (1999) The "failed state" and international law. Int Rev Red Cross 81(836):731–761

Thürer D (2009) Failing states. In: Wolfrum R (ed) Max Planck encyclopedia of public international law. Oxford University Press, Oxford (online edition)

Tomuschat C (2004) Human rights: between idealism and realism. Oxford University Press, Oxford

Tomuschat C (2009) What relationship is there between the universal legal order under the auspices of the United Nations and the EU legal order? In: Eeckhout P, Tridimas T (eds) Yearbook of European law, vol 28, pp 654–663

Trechsel S (2003) A World Court for Human Rights? Northwest Univ J Int Hum Rights 1:i–xviii, art 3
Triepel H (1923) Les rapports entre le droit interne et le droit international. Recueil des Cours I:73–121
Tullis P (2010) The U.N.'s death squad watchdog. http://www.psmag.com/legal-affairs/the-u-n-s-death-squad-watchdog-11440
Tully S (2005) A human right to access water? A critique of general comment no. 15. Neth Q Hum Rights 23(1):35–63
Tully S (2006) Flighty purposes and deeds: a rejoinder to Malcolm Langford. Neth Q Hum Rights 24(3):461–472
United Nations (UN) (2009) The millennium development goals report 2009. United Nations, New York. http://www.un.org/millenniumgoals/pdf/MDG_Report_2009_ENG.pdf
United Nations (UN) (2012) The millennium development goals report 2012. United Nations, New York. http://www.un.org/millenniumgoals/pdf/MDG%20Report%202012.pdf
United Nations Children's Fund (UNICEF), World Health Organization (WHO) (2012) Progress on drinking water and sanitation: 2012 update. http://www.unicef.org/media/files/JMPreport2012.pdf
United Nations Development Programme (UNDP) (2006) Human development report 2006: beyond scarcity – power, poverty and the global water crisis. http://hdr.undp.org/en/media/HDR06-complete.pdf
United Nations Educational, Scientific and Cultural Organization (UNESCO) (2003) The United Nations world water development report: water for people – water for life. http://unesdoc.unesco.org/images/0012/001297/129726e.pdf
United Nations Educational, Scientific and Cultural Organization (UNESCO) (2012) The 4th edition of the UN World Water Development Report (WWDR4). http://www.unesco.org/new/en/natural-sciences/environment/water/wwap/wwdr/wwdr4-2012
Unknown (2007) What price for the priceless? Implementing the right to water. Harv Law Rev 120 (4):1067–1088
van Hoof G (1984) The legal nature of economic, social and cultural rights: a rebuttal of some traditional views. In: Alston P, Tomasevski K (eds) The right to food. Martinus Nijhoff, Utrecht, pp 97–110
Vedder C (2012) Commentary on article 11 TEU. In: Vedder C, Heintschel von Heinegg W (eds) Europäisches Unionsrecht: Handkommentar. Nomos, Baden-Baden
Veigha de Cunha L (2009) Water: a human right or an economic resource? In: Lamas MR, Martinez-Cortina L, Mukherji A (eds) Water ethics: Marcelino botin water forum 2007. CRC, London, pp 97–113
Vierdag EW (1978) The legal nature of the rights granted by the international covenant on economic, social and cultural rights. Neth Yearb Int Law 9:69–105
Wälde TW (2004) ICSID 'Annulment Committee'. Transnatl Dispute Settlement 1(1). www.transnational-dispute-management.com/article.asp?key=57
Wasserstrom R (1964) Rights, human rights and racial discrimination. J Philos 61(20):628–641
Weisburd AA (1988) Customary international law: the problem of treaties. Va J Transnatl Law 21 (1):1–46
Wellman C (1995) Real rights. Oxford University Press, Oxford
Welwood W (1613) An abridgement of all sea-lawes. Man Publishing, London
Welwood W (1615) De dominio maris. Eld Publishing, London
Werner J (2009) Limits of commercial investor-state arbitration: the need for appellate review. In: Dupuy PM, Francioni F, Petersmann EU (eds) Human rights in international investment law and arbitration. Oxford University Press, Oxford, pp 115–117
Winkler IT (2010) Respect, protect, fulfil: the implementation of the human right to water in South Africa. In: Cullet P et al (eds) Water governance in motion: towards socially and environmentally sustainable water laws. Foundation Books, New Delhi, pp 415–444

Winkler IT (2012) The human right to water: significance, legal status and implications for water allocation. Hart, Oxford

Wolff HJ et al (2007) Verwaltungsrecht I, 12th edn. C H Beck, Munich

Woodhouse EJ (2003) The Guerra del Agua and the Cochabamba concession: social risk and foreign direct investment in public infrastructure. Stanford J Int Law 39(2):295–350

World Health Organization (WHO) (2012) Glass 2012 Report: UN-water global analysis and assessment of sanitation and drinking water – the challenge of extending and sustaining services. http://whqlibdoc.who.int/publications/2012/9789241503365_eng.pdf

World Health Organization (WHO), United Nations Children's Fund (UNICEF) (2006) Meeting the MDG drinking water and sanitation target: the urban and rural challenge of the decade. http://www.who.int/water_sanitation_health/monitoring/jmpfinal.pdf

World Health Organization (WHO), United Nations Children's Fund (UNICEF) (2012) A snapshot of drinking water and sanitation in Africa – 2012 update: a regional perspective based on new data from the WHO/UNICEF Joint Monitoring Programme for Water Supply and Sanitation. http://www.wssinfo.org/fileadmin/user_upload/resources/Africa-AMCOW-Snapshot-2012-English-Final.pdf

Young IM (2004) Responsibility and global labor justice. J Polit Philos 12(4):365–388

Young K (2008) The minimum core of economic and social rights: a concept in search of content. Yale J Int Law 33(1):113–175

Ziegler KS (2009) Strengthening the rule of law, but fragmenting international law: the *Kadi* decision of the ECJ from the perspective of human rights. Hum Rights Law Rev 9(2):288–305

Index

P. Thielbörger, *The Right(s) to Water*,
DOI 10.1007/978-3-642-33908-0, © Springer-Verlag Berlin Heidelberg 2014

MIX
Papier aus verantwortungsvollen Quellen
Paper from responsible sources
FSC® C105338

If you have any concerns about our products,
you can contact us on
ProductSafety@springernature.com

In case Publisher is established outside the EU,
the EU authorized representative is:
Springer Nature Customer Service Center GmbH
Europaplatz 3, 69115 Heidelberg, Germany

Printed by Libri Plureos GmbH
in Hamburg, Germany